Computerized Analysis of Mammographic Images for Detection and Characterization of Breast Cancer

Synthesis Lectures on Biomedical Engineering

Editor
John D. Enderle, *University of Connecticut*

Lectures in Biomedical Engineering will be comprised of 75- to 150-page publications on advanced and state-of-the-art topics that span the field of biomedical engineering, from the atom and molecule to large diagnostic equipment. Each lecture covers, for that topic, the fundamental principles in a unified manner, develops underlying concepts needed for sequential material, and progresses to more advanced topics. Computer software and multimedia, when appropriate and available, are included for simulation, computation, visualization and design. The authors selected to write the lectures are leading experts on the subject who have extensive background in theory, application and design.

The series is designed to meet the demands of the 21st century technology and the rapid advancements in the all-encompassing field of biomedical engineering that includes biochemical processes, biomaterials, biomechanics, bioinstrumentation, physiological modeling, biosignal processing, bioinformatics, biocomplexity, medical and molecular imaging, rehabilitation engineering, biomimetic nano-electrokinetics, biosensors, biotechnology, clinical engineering, biomedical devices, drug discovery and delivery systems, tissue engineering, proteomics, functional genomics, and molecular and cellular engineering.

Computerized Analysis of Mammographic Images for Detection and Characterization of Breast Cancer
Paola Casti, Arianna Mencattini, Marcello Salmeri, and Rangaraj M. Rangayyan
2017

Models of Horizontal Eye Movements: Part 4, A Multiscale Neuron and Muscle Fiber-Based Linear Saccade Model
Alireza Ghahari and John D. Enderle
2015

Mechanical Testing for the Biomechanics Engineer: A Practical Guide
Marnie M. Saunders
2015

Models of Horizontal Eye Movements: Part 3, A Neuron and Muscle Based Linear Saccade Model
Alireza Ghahari and John D. Enderle
2014

Digital Image Processing for Ophthalmology: Detection and Modeling of Retinal Vascular Architecture
Faraz Oloumi, Rangaraj M. Rangayyan, and Anna L. Ells
2014

Biomedical Signals and Systems
Joseph V. Tranquillo
2013

Health Care Engineering, Part II: Research and Development in the Health Care Environment
Monique Frize
2013

Health Care Engineering, Part I: Clinical Engineering and Technology Management
Monique Frize
2013

Computer-aided Detection of Architectural Distortion in Prior Mammograms of Interval Cancer
Shantanu Banik, Rangaraj M. Rangayyan, and J.E. Leo Desautels
2013

Content-based Retrieval of Medical Images: Landmarking, Indexing, and Relevance Feedback
Paulo Mazzoncini de Azevedo-Marques and Rangaraj Mandayam Rangayyan
2013

Chronobioengineering: Introduction to Biological Rhythms with Applications, Volume 1
Donald McEachron
2012

Medical Equipment Maintenance: Management and Oversight
Binseng Wang
2012

Fractal Analysis of Breast Masses in Mammograms
Thanh M. Cabral and Rangaraj M. Rangayyan
2012

Capstone Design Courses, Part II: Preparing Biomedical Engineers for the Real World
Jay R. Goldberg
2012

Ethics for Bioengineers
Monique Frize
2011

Computerized Analysis of Mammographic Images for Detection and Characterization of Breast Cancer

Paola Casti, Arianna Mencattini, Marcello Salmeri, and Rangaraj M. Rangayyan

ISBN: 978-3-031-00536-7   paperback
ISBN: 978-3-031-01664-6   ebook

DOI 10.1007/978-3-031-01664-6

A Publication in the Springer Nature series
SYNTHESIS LECTURES ON ADVANCES IN AUTOMOTIVE TECHNOLOGY

Lecture #56
Series Editor: John D. Enderle, *University of Connecticut*
Series ISSN
Print 1930-0328   Electronic 1930-0336

Computerized Analysis of Mammographic Images for Detection and Characterization of Breast Cancer

Paola Casti, Arianna Mencattini, and Marcello Salmeri
University of Rome Tor Vergata, Rome, Italy

Rangaraj M. Rangayyan
University of Calgary, Calgary, Alberta, Canada

SYNTHESIS LECTURES ON BIOMEDICAL ENGINEERING #56

ABSTRACT

The identification and interpretation of the signs of breast cancer in mammographic images from screening programs can be very difficult due to the subtle and diversified appearance of breast disease. This book presents new image processing and pattern recognition techniques for computer-aided detection and diagnosis of breast cancer in its various forms. The main goals are: (1) the identification of bilateral asymmetry as an early sign of breast disease which is not detectable by other existing approaches; and (2) the detection and classification of masses and regions of architectural distortion, as benign lesions or malignant tumors, in a unified framework that does not require accurate extraction of the contours of the lesions. The innovative aspects of the work include the design and validation of landmarking algorithms, automatic Tabár masking procedures, and various feature descriptors for quantification of similarity and for contour-independent classification of mammographic lesions. Characterization of breast tissue patterns is achieved by means of multidirectional Gabor filters. For the classification tasks, pattern recognition strategies, including Fisher linear discriminant analysis, Bayesian classifiers, support vector machines, and neural networks are applied using automatic selection of features and cross-validation techniques. Computer-aided detection of bilateral asymmetry resulted in accuracy up to 0.94, with sensitivity and specificity of 1 and 0.88, respectively. Computer-aided diagnosis of automatically detected lesions provided sensitivity of detection of malignant tumors in the range of [0.70, 0.81] at a range of falsely detected tumors of [0.82, 3.47] per image. The techniques presented in this work are effective in detecting and characterizing various mammographic signs of breast disease.

KEYWORDS

angular stationarity, architectural distortion, bilateral asymmetry, breast cancer, computer-aided diagnosis (CAD), correlation analysis, feature selection, Gabor filters, Gaussian curvature, landmarking, mammography, masses, pattern recognition, radial stationarity, structural similarity, semivariogram analysis, Tabár masking

Paola Casti dedicates this book to her son
Matteo

Contents

Preface

This work presents new image processing and pattern recognition strategies for detection and diagnosis of breast cancer with mammography. The first part of the book focuses on a computerized system for the identification of bilateral asymmetry as an early sign of tumor which is not detectable by other existing approaches. With the purpose of detecting bilateral asymmetry, novel landmarking procedures are presented to identify anatomical structures on the mammogram, including the nipple and the breast contour. Together with the pectoral muscle, they serve as landmarks for bilateral matching and, in addition, as boundary structures to confine the breast region where the abnormalities are localized. Following radiologists' criteria in interpreting mammograms, computerized Tabár masking procedures are described as a means to derive corresponding regions of the left and right mammograms for comparison and analysis. The extraction of the oriented patterns is performed to characterize the breast tissue patterns. Structural similarity or dissimilarity between paired regions on the mammograms are then quantified by means of various specifically designed measures of similarity. The feature descriptors developed for the application consist of: (1) a novel application of Moran's index to measure the angular covariance between rose diagrams related to the phase and magnitude responses of multidirectional Gabor filters; (2) features for the analysis of spatial correlation of pixel values with respect to the nipple position; and (3) spherical semivariogram descriptors and new correlation-based structural similarity indices in the spatial and complex wavelet domains.

In the second part of the book, the development of a unified and comprehensive computerized system for detection and diagnosis of breast cancer is presented, in which contour-independent diagnosis of malignant tumors, including masses and regions of architectural distortion, is performed on automatically detected suspicious focal areas. Analysis of the gradient vector field via the eigenvalues of the Hessian is performed to identify the focal areas, while a differential approach is implemented to derive features for detection of lesions. New feature descriptors are designed for quantification of 2D spatial correlation and trends over the radial and angular directions of circular regions including a lesion. They serve the purpose of contour-independent classification of candidates as benign lesions or malignant tumors. Finally, a 3D free-response receiver operating characteristic framework is introduced for evaluation of two binary categorization problems in series.

Sequential forward/backward selection and stepwise logistic regression are used for automatic selection of the various extracted features. Pattern recognition techniques, including Fisher linear discriminant analysis, Bayesian classifiers, support vector machines, and neural networks, are applied to experimental training data and the obtained models are used for the automatic classification of test data. The effectiveness of the developed systems is demonstrated through cross-validation techniques such as leave-one-patient-out and k-fold analysis. Two public databases, Digital Database for Screening Mammography (DDSM) and Mammographic Image Analysis Society (MIAS) database, together with a private database of full-field digital mammograms (FFDMs) from San Paolo Hospital of Bari, Italy, are used for the analysis. Multiple comparisons of the results achieved in this work with the results reported in previous research work are reported.

Computer-aided detection (CADe) of bilateral asymmetry has not been studied adequately. There is increasing interest in this area, as indicated by the appearance of publications addressing the problem. The performance of the methods developed in this work for the detection of bilateral asymmetry resulted in accuracy up to 0.94, with sensitivity and specificity of 1 and 0.88, respectively; the obtained results are better than the results reported in other works in the scientific literature and are expected to improve the performance of techniques for mammography and breast cancer.

Computer-aided diagnosis (CADx) of mammographic lesions, in particular of automatically detected masses and regions of architectural distortion, is an important but yet-to-be addressed task that can facilitate accurate interpretation of mammograms and reduce unnecessary breast biopsies. Some of the related subproblems have been addressed by researchers as independent tasks. The integration of the various aspects of detection and classification of mammographic lesions pose a new challenge to be addressed, which demands the design of a unified CADe/CADx system. Moreover, the presence of tumors with obscured or ill-defined margins, for which the existing approaches based on accurate segmentation of the lesions are prone to fail, has motivated the contour-independent approach presented in this work. The results obtained with the CADe/CADx system indicate sensitivity of detection of malignant tumors in the range of [0.70–0.81] at a range of falsely detected tumors of [0.82–3.47] per image. The results obtained with FFDMs, in particular, compared favorably with the performance of the existing commercial systems for the automatic detection of masses, which is a simpler problem than integrating detection and classification of lesions. The methods presented in this work are expected to improve the scope and performance of CADe/CADx systems for breast cancer.

Acknowledgments

The research work presented in this book would not have been possible without the valuable support of many individuals and organizations.

We thank the Diagnostic Radiology Unit, San Paolo Hospital of Bari, Italy, in particular Dr. Antonietta Ancona, Dr. Fabio F. Mangieri, and Dr. Maria Luisa Pepe from the Integrated Operational Unit of Diagnostic Imaging, ASL, Taranto, Italy, and Prof. Maria Grazia Raguso from the Department of Mathematics, University of Bari, Italy, for providing the digital mammograms used in this work, the ground truth annotations, and the support needed for validation of results.

We thank J. Pont and E. Pérez from the Department of Radiology of the Girona University Hospital "Dr. Josep Trueta" (Spain) and E.R.E. Denton from the Department of Breast Imaging of the Norwich and Norfolk University Hospital (UK) for providing the BI-RADS classification of the MIAS database used in this work. We also thank Dr. Ricardo J. Ferrari from Departamento de Computação, Universidade Federal de São Carlos, São Paulo, Brasil, and his colleagues for having made available additional ground-truth contours of the MIAS database.

We gratefully thank the Department of Electronics Engineering of the University of Rome Tor Vergata, Italy, including the affiliated professors, researchers, technical and administrative personnel, and students who supported our work and enriched it with their unique contributions.

Some of the materials and illustrations have been reproduced, with permission, from the associated organizations, from our publications listed below.

1. P. Casti, A. Mencattini, M. Salmeri, A. Ancona, F. Mangieri, M.L. Pepe, and R.M. Rangayyan, Contour-independent detection and classification of mammographic lesions, *Biomed. Signal Process. Control*, 25:165–177, 2016. © Elsevier.

2. P. Casti, A. Mencattini, M. Salmeri, and R.M. Rangayyan, Analysis of structural similarity in mammograms for detection of bilateral asymmetry, *IEEE Trans. Med. Imag.*, 34(2), 662–671, 2015. © IEEE.

3. P. Casti, A. Mencattini, M. Salmeri, A. Ancona, F. Mangieri, M.L. Pepe, and R.M. Rangayyan, Estimation of the breast skin-line in mammograms using multidirectional Gabor filters, *Comput. Biol. Med.*, 43(11), 1870–1881, 2013. © Elsevier.

4. P. Casti, A. Mencattini, M. Salmeri, A. Ancona, F. Mangieri, M.L. Pepe, and R.M. Rangayyan, Automatic detection of the nipple in screen-film and full-field digital mam-

mograms using a novel Hessian-based method, *J. Dig. Imag.*, 26(5), 948–957, 2013. © Springer.

5. P. Casti, A. Mencattini, M. Salmeri, and R.M. Rangayyan, Masking procedures and measures of angular similarity for detection of bilateral asymmetry in mammograms, *4th IEEE Conference on e-Health and Bioengineering (EHB)*, Iasi, Romania, 2013. © IEEE.

6. P. Casti, A. Mencattini, M. Salmeri, A. Ancona, F. Mangieri, M.L. Pepe, and R.M. Rangayyan, Design and analysis of contour-independent features for classification of mammographic lesions, *Proc. of the 4th IEEE Conference on e-Health and Bioengineering (EHB)*, Iasi, Romania, 2013. © IEEE.

7. P. Casti, A. Mencattini, M. Salmeri, F. Mangieri, and R.M. Rangayyan, Measures of radial correlation and radial trends for classification of breast masses in mammograms, *Proc. of the 35th IEEE Annual International Conference of the IEEE Engineering in Medicine and Biology Society (EMBC)*, Osaka, Japan, 2013. © IEEE.

8. A. Mencattini, M. Salmeri, P. Casti, and M.L. Pepe, Local active contour models and Gabor wavelets for an optimal breast region segmentation, *Proc. of the 26th International Congress and Exhibition: Computer Assisted Radiology and Surgery (CARS)*, Pisa, Italy, 2012. © Springer.

We thank our families, relatives, and friends for their loving support, encouragement, and understanding, and for making our lives meaningful.

Paola Casti, Arianna Mencattini, Marcello Salmeri, and Rangaraj M. Rangayyan
May 2017

CHAPTER 1

Introduction

1.1 BREAST CANCER AND MAMMOGRAPHY

1.1.1 BREAST CANCER STATISTICS

With 1.67 million new cases in 2012, breast cancer represents 25% of all diagnoses of cancer cases worldwide. The estimates for 2015 indicate 1.70 million of new cases [64]. Among women, breast cancer is the most frequent form of cancer and it is second only to lung cancer as the most frequent cause of cancer death worldwide. According to the ISTAT [66], breast cancer is both the most frequent form of cancer and the most common cause of cancer death in all age groups of Italian women: it accounts for 28% of deaths among young women, 21% among adults, and 14% among women aged more than 70 years.

When breast cancer is diagnosed at an early stage, the prognosis for the patient is favorable and surgery can be resolutive even if limited to the lesion and its surrounding tissue. At higher stages of the disease, the surgical procedure may need to be followed by radiation therapy or, in some cases, chemotherapy, which have the purpose of protecting the remaining glandular tissue from the risk of local recurrence of cancer [137]. However, 25% of women with breast cancer present are diagnosed with advanced forms of the disease that need to be treated with more aggressive treatments consisting of the removal of the entire breast, named radical mastectomy, and involving chemotherapy before surgery, followed by additional chemotherapy and radiation; in such cases, the chances of survival are drastically reduced [84].

1.1.2 MAMMOGRAPHY SCREENING PROGRAMS

Through the years, researchers have learned that early diagnosis is critical to cure breast cancer and that access to screening tools by women facilitates efficient treatments for most patients [40]. If the cancer is detected at an early stage, in fact, more treatment options are available and the patient's life can be saved. At present, the best radiographic method for detecting breast cancer is mammography. Therefore, cancer screening is best provided by the combination of mammography and clinical breast examination, performed at standard intervals. The majority of European countries adopted national or regional mammography screening programs, which consist of periodical mammographic examinations of asymptomatic women aged 50–69 years with a 2-year screening interval [112]. The accumulated evidence [144] suggests that a reduction of 20% in breast cancer mortality can be achieved by organized mammography programs. A recent study of the IMPACT Working Group [51] has evaluated the effects of the national mammographic screening programs on the incidence of breast cancers at advanced stages in Italy, showing a significant and stable reduction in breast cancer-related specific mortality in the range of 20–30%.

1.1.3   THE MAMMOGRAPHIC EXAMINATION

The beginning of mammographic examinations dates back to 1913, when the German surgeon Albert Salomon obtained the first X-ray image of breast tissue removed from cancer patients [115]. Almost 20 years later, the American radiologist Stafford L. Warren performed the first in vivo mammography by means of an apparatus for conventional X-ray imaging. Since then, physicians all over the world have contributed to improvements of the technique, and today, mammography has become the recognized technique for detection of breast cancer worldwide [115].

Mammography consists of an X-ray examination of the breast made with a specific X-ray equipment that is capable of finding tumors too small to be palpable. The output is a shadowgram of the breast which is recorded by an image receptor and that results from the attenuation of X-rays along paths passing through the glandular structures. The projected structures are magnified onto the image receptor due to the spreading of the X-rays from the source, while the differential X-ray attenuation among the various tissue structures is responsible for the image contrast. The magnification of the internal structures of the breast and the obtained contrast enable, in the presence of pathological processes, the identification of alteration of the mammary gland. With this purpose, the spatial resolution of the mammographic system can be as small as 10 μm in order to discriminate fine details and low-energy X-rays in the range of 24–32 keV are used to obtain good contrast. Limits to the image quality are posed by the restrictions on the radiation dose absorbed by the patient, which should be less than 3 mGy, and by an overall random fluctuation, referred to as mottle or noise. Such limitations, together with the superimposition of tissue during projection, contribute to make the mammographic interpretation difficult even by experienced radiologists [14].

Screen-film mammography (SFM) and full-field digital mammography (FFDM) are the two modalities for performing mammographic examinations nowadays. In SFM, which is the conventional analog system, the X-ray photons are converted to light by a phosphorescent screen; the light image is captured by a film, which is then developed for interpretation by the radiologist. Digital images can be produced by digitization after the film is processed. In FFDM, the X-ray photons are converted by solid-state detectors into electronic signals that can be displayed directly on a high-resolution monitor. SFM had been the modality of choice for screening programs; however, FFDM is replacing conventional film imaging systems, mainly due to increased quality of images and a wider dynamic range, in addition to the benefits of digital technology in data transmission, retrieval, display, and storage [62].

During screening mammography, the left and right breasts of a woman are imaged separately. Two views are obtained by compressing each breast along different directions of projections to cover the majority of fibroglandular structures. The craniocaudal (CC) view is acquired by a vertical projection of the breast, which is compressed in the head-to-toe direction. The mediolateral-oblique (MLO) view is taken by compressing the breast from the middle of the chest to the outside of the body in an oblique direction. In Figs. 1.1a and b, two normal mammograms of the right breast of a patient in the CC and MLO projections, respectively, are shown. The brighter areas at the center of the images correspond to the fibroglandular tissue. The nipple is in profile and visible in both views, while the pectoral muscle is visible only in the MLO view. The mammograms were acquired at the San Paolo Hospital of Bari, Italy. More information on the related database and how the images were obtained are provided in Section 2.1.1.

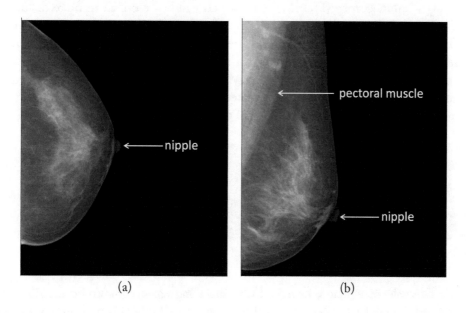

(a)　　　　　　　　　　　　　(b)

Figure 1.1: (a) The craniocaudal (CC) view and (b) the mediolateral-oblique (MLO) view of the right breast of a patient. The mammograms show normal breast parenchyma with normal radiographic density. The FFDM images were acquired at the San Paolo Hospital of Bari, Italy (see Section 2.1.1 for more details).

1.1.4 MAMMOGRAPHIC SIGNS OF BREAST DISEASE

The Breast Imaging Reporting and Data System (BI-RADS) of the American College of Radiology (ACR) [39], which is the standard reference for reporting mammographic results, describes four categories of breast abnormalities that can indicate breast cancer: masses, calcifications, architectural distortions, and bilateral asymmetry.

Masses

A mass is defined as a space occupying lesion seen in two different projections [39]. When a potential mass is seen in a single view it is named "density" until its nature is confirmed. Masses usually appear as areas brighter than the surrounding tissue due to increased attenuation of X-rays if they possess higher density, but they can also result in equal density (isodense) or less density (hypodense) regions on the mammogram. There are also fat containing masses which appear as radiolucent regions. The margins of masses are classified as circumscribed, microlobulated, obscured, indistinct, or spiculated [39]. Examples of masses from each category of margins are illustrated in Fig. 1.2. The regions of interest (ROIs) are from the Digital Database for Screening Mammography (DDSM) [61], which will be described in more detail in Section 2.1.3. Different shapes of masses are also possible: round, oval, lobular, or irregular [39].

When a suspicious area is detected on the mammogram by the radiologist, its nature is further investigated by means of a histological examination of a biopsy sample of the lesion in order to determine whether the area corresponds to a malignant tumor or a benign lesion. Tumors as small as 2 mm in diameter can be detected on the mammogram but are also the most difficult to identify, especially in the presence of dense fibroglandular tissue [69].

| (a) | (b) | (c) | (d) | (e) |

Figure 1.2: Examples of regions of interest (ROIs) including a mass with (a) circumscribed, (b) microlobulated, (c) obscured, (d) indistinct, or (e) spiculated margins. (a,c) Benign lesions. (b,d,e) Malignant Tumors. The ROIs are extracted from the Digital Database for Screening Mammography (DDSM) [61]; see Section 2.1.3 for more details.

Calcifications

Calcifications are small and bright spots on the mammogram due to the deposition of calcium within the breast parenchyma. They are characterized in terms of size, morphology, number, and distribution [39]. The presence of a cluster of calcifications is associated with an increased risk of malignancy, but benign clusters may also occur. The detection of calcifications is limited by the signal-to-noise ratio. The size varies from 100 μm to 1 mm. However, the typical size of calcifications that is detectable by conventional mammography is around 200 μm. The detection of smaller calcifications requires geometric magnification techniques, which are performed as part of a diagnostic mammogram workup for patients suspected to have breast disease but not during screening mammography [5]. Examples of calcifications from the DDSM are shown in Fig. 1.3.

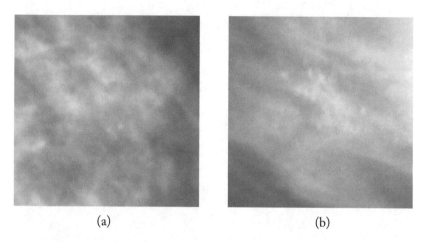

(a)   (b)

Figure 1.3: Examples of (a) benign and (b) malignant clusters of calcifications from the DDSM [61].

Architectural Distortion

When the normal architecture of the breast parenchyma is distorted in the absence of visible masses, it is defined as architectural distortion. It is usually characterized by the presence of spiculations radiating from a point, but it can have a more subtle appearance manifested by focal retraction or distortion at the edge of the parenchyma [39]. Two examples of regions of architectural distortion from the DDSM [61] are illustrated in Figs. 1.4a and b.

Bilateral Asymmetry

Radiologists perform comparative studies of the left and right mammograms of a given patient to prevent missing signs of breast disease. When a greater area of tissue with fibroglandular density is detected in a mammogram relative to the corresponding region in the contralateral breast, it is reported as an asymmetric finding, either local or global [39]. The presence of a greater area of tissue with fibroglandular density when judged relative to the contralateral breast defines a class of mammographic lesions denoted as bilateral asymmetry [39]. The condition of asymmetry is reported as global if the observed differences in areas of fibroglandular tissue are extensive, or focal if the difference in fibroglandular tissue density is confined to a small region but lacks the conspicuity of a mass [39]. Examples of two pairs of focal and global bilateral asymmetry from the DDSM [61] are illustrated in Fig. 1.5a and b, respectively. The identification of all asymmetric findings in a given pair of mammograms is important, because they may be the only clue to breast disease that is detectable on standard mammographic projections, especially when

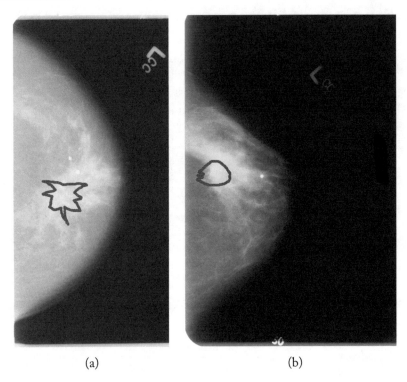

(a)   (b)

Figure 1.4: Examples of mammograms with regions of architectural distortion from the DDSM [61]. The red contours indicate the regions outlined by the radiologist.

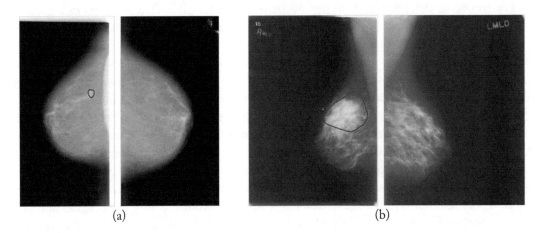

(a)   (b)

Figure 1.5: Examples of two pairs of (a) focal and (b) global bilateral asymmetry from the DDSM [61].

masses, microcalcifications, and/or architectural distortion are not visible [134]. Bilateral asymmetry has proved to be an indicator of increased risk of developing breast cancer [59, 133, 170], stressing the importance of special surveillance and follow-up observations of the patients to establish the nature of the asymmetry present. Asymmetric findings on mammograms may indicate a developing or underlying mass. They can be subtle in presentation and hence overlooked or misinterpreted by radiologists. The difficulty with the detection of asymmetry arises because the bilateral anomalies caused by a developing or underlying pathological process need to be differentiated from the physiological differences between the two breasts and distortions due to projection artefacts. These confounding factors and subtlety in presentation can cause overlooking or misinterpretation, even by experienced radiologists [88]. Clinical studies have reported that asymmetry accounts for 3–9% of breast cancer cases incorrectly reported by radiologists as showing no evidence of a tumor [15]. Evidence also suggests that asymmetric distribution of fibroglandular density is a common source of false-positive (FP) diagnosis [154].

1.1.5 BI-RADS MAMMOGRAPHIC DENSITY CATEGORIES

Mammograms exhibit differences in terms of the type of breast tissue composition. Due to the variable proportion of fatty and fibroglandular tissues in the breast composition mammographic images are difficult to interpret by radiologists. Fibroglandular tissue is composed by the stroma that forms the connective components of the breast, by the glandular component that represents its functional part, and by the breast ducts that correspond to the mammary parenchyma. Fatty and fibroglandular tissues have different X-ray attenuation coefficients. In particular, regions of fat appear darker than the fibroglandular components on the mammogram. Fibroglandular regions appear as brighter regions and are referred to as "mammographic density." It is well known that a strong correlation exists between the presence of relatively large regions of density in mammograms and the risk of developing breast cancer in the near term [161]. Detecting breast cancer in mammograms of dense breasts is more difficult due to the superimposition of projected tissues that may obscure small tumors. As a consequence, the accuracy of mammography is inversely correlated with density. It is then important to determine the category of density of a subject in order to have an indication of the detection capability of the examination [82]. The BI-RADS [39] lexicon defines four density classes (B-I to B-IV) and establishes the corresponding effects on the diagnostic accuracy. The accuracy of mammography to detect suspicious lesions decreases for types III and IV as follows.

B-I: The breast is almost entirely fat and the accuracy of mammography is very high.

B-II: The breast has scattered fibroglandular densities and the accuracy of mammography is high.

B-III: The breast is heterogeneously dense and the accuracy of mammography is limited.

B-IV: The breast is extremely dense and the accuracy of mammography is limited.

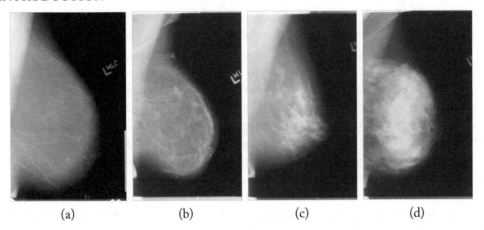

(a)   (b)   (c)   (d)

Figure 1.6: Examples of mammograms from the DDSM [61] for the BI-RADS mammographic density categories (a) B-I, (b) B-II, (c) B-III, and (d) B-IV.

1.1.6   TABÁR MASKING

The perception of subtle radiographic abnormalities in breast cancer screening can be improved by the use of a systematic approach to the analysis of mammograms, aimed at reducing false-positive rates (FPRs) and maintaining high levels of sensitivity. A complete mammographic study requires side-by-side viewing of corresponding areas of both breasts, whose practical realization should be strengthened by the technique of masking, as described by Tabár [145], ensuring that all regions of the breasts are viewed and compared in detail with the contralateral regions.

An exhaustive Tabár masking process would require at least four different types of analysis for each pair of views, performed with stepwise movements: horizontal and oblique masking of the MLO views, both in the cranial and caudal directions; horizontal and vertical masking of the CC views, the former in the medial and lateral directions and the latter in the proximal and distal directions. Particular attention should be given to the so-called "forbidden areas," specific areas where the majority of breast cancers are found in the early phase [145]:

(a) *medial-half area*: the medial half of the breast on CC projections;

(b) *retroglandular area*: the retroglandular space on CC projections;

(c) *milky area*: the region parallel with the edge of the pectoral muscle on MLO projections; and

(d) *retroareolar area*: the retroareolar region on MLO projections.

The "forbidden areas" on mammograms are illustrated in Figs. 1.7 and 1.8 together with possible masking procedures. During the masking procedures, different regions of the mammograms are compared by a radiologist with step-by-step movements. At each step, the areas under investigation can be matched singularly, changing stepwise the analyzed regions (*Stepwise Tabár masking*), or gradually, increasing the size of the paired observation windows (*Incremental Tabár masking*). The first approach enhances the perception of focal anomalies, whereas the latter allows a better understanding of global changes in the breast parenchyma.

1.1.7   DRAWBACKS AND LIMITATIONS OF MAMMOGRAPHY

Although mammography is the most widely used screening modality with solid evidence of benefit for women, some authors have been stressing its limitations, including false-negative (FN) and FP outcomes, overdiagnosis, and overtreatment of patients [2, 97]. Mammographic interpretation is a difficult task: signs of breast cancer can be very subtle and are often obscured by normal fibroglandular breast tissue with which these signs have many features in common, making their visual detection and analysis difficult. The accuracy of interpretation of screening mammograms, in particular, is affected by several factors, such as image quality and the radiologists' level of expertise. The rate of malignant cases missed by radiologists in the past few years has been reported to be 10–30% [88]. Additional reasons include the low prevalence of the disease in a screening population and the large number of mammograms that radiologists need to assess every day. To overcome such limitations and to improve radiologists' performance in interpreting mammograms, double reading, which consists of having two radiologists interpreting each case independently, has been introduced [54]. The alternative to double reading in the current screening practice consists of the use of computerized systems as second readers.

1.2   COMPUTER-AIDED DETECTION AND DIAGNOSIS WITH MAMMOGRAPHY

1.2.1   THE ROLE OF CAD AS A SECOND READER

Computer-aided detection and diagnosis (CAD) techniques and systems involve the use of computer algorithms to detect patterns in images associated with signs of disease. In mammography, they can support radiologists in the role of a second reader, prompting the radiologists to review areas in a mammogram deemed to be suspicious (computer-aided detection, or CADe) and distinguishing between a lesion that is decidedly negative on a mammogram as opposed to one that needs regular monitoring or requires a biopsy (computer-aided diagnosis, or CADx).

For an understanding of the limits and potential of CAD of breast cancer, it is of interest to report what Alan Turing, the father of theoretical computer science and artificial intelligence, said about computing machinery and intelligence:

> *"I would say that fair play must be given to the machine. Instead of it giving no answer we could arrange that it gives occasional wrong answers. But the human mathematician*

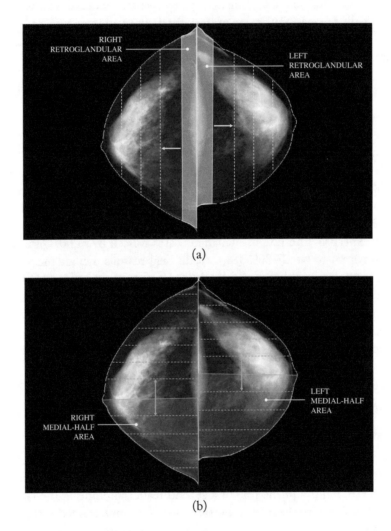

(a)

(b)

Figure 1.7: Masking procedures and "forbidden areas" (shown with labels) on mammograms in CC projections.

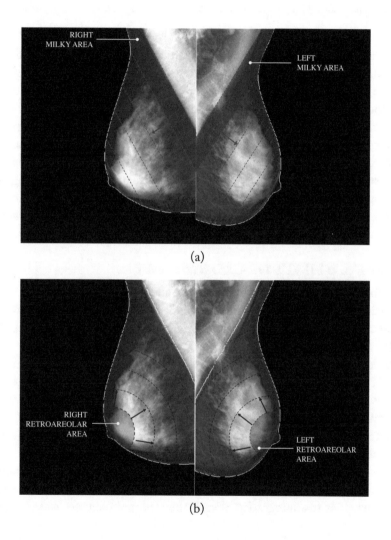

(a)

(b)

Figure 1.8: Masking procedures and "forbidden areas" (shown with labels) on mammograms in MLO projections.

would likewise make blunders when trying out new techniques. In other words then, if a machine is expected to be infallible, it cannot also be intelligent. There are several mathematical theorems which say almost exactly that. But these theorems say nothing about how much intelligence may be displayed if a machine makes no pretence at infallibility."

Alan Turing 1912–1954

Beside the potentials of artificial intelligence, Turing stressed the importance of its limits. Such limits, today, are at the basis of a correct fruition of computerized systems. Accurate quantification of the performance of CAD systems in solving real problems is crucial for efficient physician computer interaction. However, given the limits and benefits of both computer and human vision, the key for improved levels of sensitivity and specificity of diagnostic tests lies in the integration of automated approaches for quantitative analysis with human intuition. The final performance of a CAD system in mammographic reading should correspond to the performance achieved by the radiologist when interpreting mammograms by using the computer output as a second opinion, not by the performance of the CAD system itself. In this way, even if the accuracy levels of CAD systems do not surpass the levels achievable by radiologists, it is their interaction with the radiologist that determines the final benefits of CAD.

1.2.2 CLINICAL UTILITY OF CAD SYSTEMS

There are commercially available CAD systems for mammography whose benefits in a screening or diagnostic environment have been evaluated. Two of the most widely available are the ImageChecker system (Hologic, Inc., Bedford, MA) and the SecondLook system (iCAD, Inc., Nashua, NH).

The first large reported study on the effects of CAD systems in a screening environment was performed by Freer and Ulissey [52]; over a 12-month period the radiologists' performance using CAD was analyzed with respect to the radiologists' performance without CAD, showing an increase of 19.5% in the detected cancers. The study also reported a clinically significant ability of CAD systems in detecting tumors at an early stage (0 and I), which is critically important for saving women's lives. A consequent drawback was an increase of 19% in the number of recalled patients. However, the overall positive-predictive value for biopsy remained unchanged. Subsequent studies [11, 76, 101] confirmed the benefits in breast cancer detection due to the use of CAD systems for mammography, even if the reported increases were lower and ranged from 4.7–16.1%.

The next level of evaluation of CAD systems consists of determining the performance of radiologists using CAD in order to evaluate the effectiveness of CAD as an alternative to double reading. The interpretation of each case by two radiologists separately has been shown to increase the rate of cancer detection by 4–14%, but there are situations where double reading is not practiced [54]. The study by Ciatto et al. [34] estimated an increase of sensitivity of detection of interval cancers with prior negative screening mammograms of around 0.9% by introducing CAD-assisted reading with respect to double reading. A subsequent large study by

Gilbert et al. [54] found evidence on the equivalence of single reading with CAD and double reading. The economic and social benefits of the two alternative screening approaches have been analyzed in a recent study by Sato et al. [132], which found the use of CAD-assisted reading more cost-effective than double reading. However, the benefits due to the use of CAD systems in mammography are still a matter of debate and need further investigation. The same authors as cited above [11, 34, 52, 76, 101], in fact, have also reported the additional recall in 8–35% of cases produced by CAD and some results are still controversial. The study by Fenton et al. [46], for example, concluded that the increased rate of biopsy caused by CAD technology is not correlated with improved detection of breast cancer. The results are seen as the major limitation of CAD systems for mammography because of the increased costs of unnecessary followup examinations, resulting in additional anxiety for patients. Moreover, although several CAD techniques are effective in detecting masses and calcifications, the results are still less favorable for bilateral asymmetry, architectural distortion, and masses with obscured or ill-defined margins. Research is still in progress to overcome such limitations and also to move from CADe to CADx. New solutions can be found via image processing and computer vision techniques to be integrated into routine clinical practice and to improve the existing systems.

1.2.3   STATISTICAL EVALUATION OF DIAGNOSTIC PERFORMANCE

The diagnostic performance of any computerized system for mammography is assessed by comparing the result of the prediction test with a gold standard test. In fact, the gold standard test defines unequivocally the presence or absence of disease, as does the radiologist by reporting the presence of abnormalities or the biopsy result by indicating the benign or malignant nature of a lesion. The result of a prediction test for a two-class problem, as a diagnostic test, is negative, if it does not indicate the presence of the disease, or it is positive, when it indicates instead that the disease is present. For a given instance, the CAD system can give four possible outcomes with reference to the gold standard test.

TP (True Positive): the CAD system makes a positive prediction and the instance is actually positive.

TN (True Negative): the CAD system makes a negative prediction and the instance is actually negative.

FP (False Positive): the CAD system makes a positive prediction but the instance is actually negative (type I error).

FN (False Negative): the CAD system makes a negative prediction but the instance is actually positive (type II error).

The relative frequencies of the correct results obtained on the set of all instances define the sensitivity, or TPR, and the specificity, or true-negative rate (TNR), of the system as:

$$TPR = \frac{\#TP}{(\#TP + \#FN)} \tag{1.1}$$

and

$$TNR = \frac{\#TN}{(\#TN + \#FP)}, \tag{1.2}$$

which give an indication on how reliable the system is in making positive and negative identifications, respectively. In addition, the relative frequencies of the obtained incorrect results are quantified by means of the false-positive rate (FPR) and false-negative rate (FNR) of the system as:

$$FPR = \frac{\#FP}{(\#FP + \#TN)} \tag{1.3}$$

and

$$FNR = \frac{\#FN}{(\#FN + \#TP)}. \tag{1.4}$$

A global index of reliability is given by the accuracy of the system, which is defined as

$$Accuracy = \frac{\#TN + \#TP}{(\#TN + \#FP + \#TP + \#FN)}. \tag{1.5}$$

However, the values of TPR, TNR, FPR, FNR, and accuracy provide a static representation of the diagnostic test that does not consider the strength with which each instance belongs to one of the two classes. In fact, for every test, the calculated values of the described performance indices vary based on the particular cutoff value chosen to distinguish normal and abnormal results. This aspect is shown in Fig. 1.9, where increasing the cutoff level would make the test more specific but less sensitive: decreasing the number of FP instances and increasing the number of FN instances. Similarly, lowering the cutoff value would increase sensitivity while decreasing specificity: lowering the number of FP instances and decreasing the FN instances.

The decision about what cutoff to use in calling a test abnormal corresponds to a decision about whether it is better to tolerate FP instances (people without the disease inappropriately classified as diseased) or FN instances (missed cases with the disease). The choice of the cutoff depends on the disease in question and on the purpose of test. A good screening system requires a high TPR to detect all the possible lesions, giving the possibility of treating them early, and a high TNR to avoid useless and expensive further investigation of cases without the disease. This may not be achievable in reality. For the above-mentioned motivations, the best way to characterize a test is by extending the analysis of the diagnostic performance of a test by varying the cutoff, or threshold. The typical way to show this relationship is to plot the pairs (TPR, $1 - TNR$) for the entire range of possible threshold values. The resulting curve, known as the receiver operating characteristics (ROC) curve, which was originally described by researchers investigating

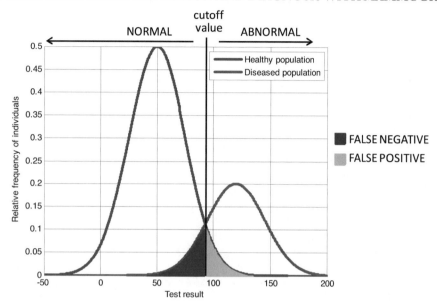

Figure 1.9: Example of a frequency distribution of a diagnostic test result in healthy and diseased individuals.

methods of electromagnetic-signal detection [113] during World War II and later applied to the field of psychology, facilitates improved analysis of the classification performance of a diagnostic method. A perfect diagnostic system has a ROC curve defined by the pairs $(0, 0)$, $(0, 1)$, and $(1, 1)$, whereas a random guess system corresponds to the diagonal line that goes from $(0, 0)$ to $(1, 1)$ [121]. The area under the ROC curve (AUC), also named A_z when estimated with a binormal model [140], ranges from zero to one and provides a measure of the system as ability to discriminate between actual positive cases and actual negative ones. AUC = 0.5 corresponds to random guess and AUC = 1 indicates an ideal diagnostic test with perfect separation between the positive and the negative classes. An example ROC curve with AUC = 0.79 is illustrated in Fig. 1.10.

When discrete and countable abnormalities (such as masses) need to be detected, it is important to establish the values of TPR obtained by the diagnostic system against the number of FPs per image (FPpI). This extends the ROC curve to a form known as the free-response receiver operating characteristic curve (FROC) [121].

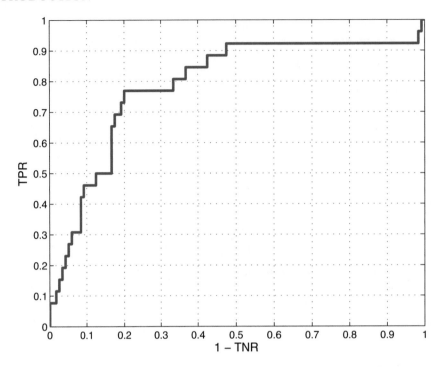

Figure 1.10: Receiver operating characteristic curve for a hypothetical diagnostic test with AUC = 0.79.

1.3 SCOPE AND ORGANIZATION OF THE BOOK

1.3.1 AIMS OF THE WORK

In the work presented in this book, four main tasks within the context of mammography and breast cancer have been addressed by means of image processing and pattern recognition techniques.

1. Design of novel landmarking algorithms for the extraction of the breast-skin line and the detection of the nipple. Reference anatomical structures, or landmarks, on mammograms, i.e., the nipple, breast-skin line, and pectoral muscle, are used in this work for effective matching between corresponding regions of the left and right breasts of a patient with the aim of performing automatic Tabár masking procedures and quantitative bilateral comparisons between pairs of mammograms.

2. Development and validation of a CAD system for the identification of bilateral asymmetry in mammograms as an early sign of breast cancer. The procedure is based on the analysis of the structural similarity or dissimilarity between paired mammographic regions. The

methods should improve the diagnostic sensitivity of CAD systems for mammography by providing clues about the presence of breast cancer which are not detected by other existing approaches.

3. Design of features for classification of mammographic regions as benign lesions or malignant tumors without relying on accurate extraction of the contours of the lesions. This is important, especially for masses or regions of architectural distortion with obscured or ill-defined margins, for which the other available approaches based on the segmentation of the lesions are prone to fail.

4. Development and validation of a novel comprehensive and multistage CADe/CADx system for automatic detection and diagnosis of malignant tumors. The task is addressed in a realistic scenario of a three-class environment, i.e., in the presence of normal parenchymal tissue, benign lesions, and malignant tumors, including masses and regions of architectural distortion. Integrated systems for mammography are expected to reduce the recall rate in screening mammography.

1.3.2   OVERVIEW

Chapter 2 documents the characteristics of the three databases of mammograms used in this work, including SFMs and FFDMs, as well as the validation strategy implemented.

Chapter 3 presents a technique for the extraction of the directional components of the breast used as a fundamental step in the analysis of mammograms.

Chapter 4 presents algorithms for automatic detection of anatomical reference structures, or landmarks, on the mammogram, including an algorithm for detection of the pectoral muscle and new methods for extraction of the breast-skin line and for detection of the nipple.

Chapter 5 contains novel methods for quantification of structural similarity or dissimilarity between the right and left mammograms of a patient as part of a CAD system for detection of bilateral asymmetry.

Chapter 6 provides details on the design process of contour-independent features for classification of masses as benign lesions or malignant tumors.

Chapter 7 focuses on the various original steps of analysis developed to design an integrated CADe/CADx system for detection and diagnosis of malignant tumors, including masses and regions of architectural distortion.

CHAPTER 2

Experimental Setup and Databases of Mammograms

All methods and procedures for the automatic detection and characterization of breast cancer designed and presented in this book have been tested using sets of digitized SFMs publicly available and/or a private database of FFDMs. Also, validation by expert radiologists was needed. Each of the various tasks addressed, in particular, has required the selection of specific sets of mammograms for the following reasons: (1) optimization of the cross-validation procedures with reference to specific lesions, (2) availability of radiologists' annotations, and (3) implementation of comparative analysis of the results obtained in this study with the results reported by previous work on the same sets of images. The use of mammograms acquired with different technologies for testing the methods proposed in this research work was crucial to prove the robustness of the developed systems and to guarantee their applicability in real practice. In addition, comparisons of the performance achieved with the developed methods with the performance levels reported by other research groups on publicly available images are necessary to verify eventual improvements in the state of the art of CAD systems for mammography.

2.1 DATABASES OF MAMMOGRAMS

2.1.1 FFDM DATABASE

The FFDM database is composed of a set of 194 FFDMs acquired at the Diagnostic Radiology Unit, San Paolo Hospital of Bari, Italy, with a GE Senographe 2000D ADS 17.3 from GE Medical Systems. Each study in the Digital Imaging and Communications in Medicine (DICOM) [103] format includes the four standard views: two CC and two MLO views. However, some of the studies only included the CC views of the patient. The dataset contains 102 CC views and 92 MLO views. The images have a spatial resolution of 94 μm/pixel and a pixel resolution of 12 bits per pixel (bpp). All the studies were directly anonymized in their original DICOM file format and then the images were converted into the Portable Grayscale Map (PGM) format by means of a proprietary conversion program. Informed consent for anonymous use of sensitive data for scientific purposes was obtained from all patients.

2.1.2   MIAS DATABASE

The Mammographic Image Analysis Society (MIAS) database [141] includes 320 films taken from the UK National Breast Screening Program and digitized originally to 50 μm/pixel and a gray-scale resolution equal to 8 bpp, with a Joice-Loebl scanning microdensitometer. Mammograms are available online via the Pilot European Image Processing Archive (PEIPA) of the University of Essex, at:

http://peipa.essex.ac.uk/info/mias.html.

The database is composed of left and right MLO mammograms from 161 patients and contains 322 digitized SFMs, including radiologists' reports on any abnormalities that may be present, together with the marking on the corresponding locations (center and radius of the lesion) and assessment of malignancy provided by biopsy. The original database was reduced to a spatial resolution of 200 μm/pixel edge and padded/clipped so that each image is represented by a 1024×1024 matrix and the mammogram is centered in the matrix (in the mini-MIAS database).

2.1.3   DDSM

The Digital Database for Screening Mammography (DDSM) [61] is a publicly available database of mammograms containing a total of 2,620 images, which were obtained from the Massachusetts General Hospital, Wake Forest University School of Medicine, Sacred Heart Hospital, and Washington University School of Medicine. The images were digitized using four different scanners: the DBA M2100 ImageClear, Howtek 960, the Lumisys 200 Laser, and the Howtek MultiRad850, which have sampling rates equal to 42, 43.5, and 50 μm/pixel, respectively, and gray-scale resolution equal to 12 bpp for the Howtek 960, Lumisys 200 Laser, and Howtek MultiRad850, and 16 bpp for the DBA M2100 ImageClear. Each study contains two MLO and two CC views, the BI-RADS density category of the images, as well as proven ground truth of the lesions present and the corresponding information on the margins, shapes, and subtlety. All the contours of the lesions were delineated by expert radiologists and are available for analysis. More detailed information is available online at

http://marathon.csee.usf.edu/Mammography/Database.html.

Table 2.1 summarizes the characteristics of the three databases used in this work, while Fig. 2.1 shows examples of mammograms.

2.2   VALIDATION STRATEGY

Validation of the various algorithms and methods presented in this book was performed by means of proved ground truth consisting of biopsy results and radiologists' manual annotations of landmarks and lesions. The annotations, which were not publicly available, were provided

Table 2.1: Databases of mammograms

Database	FFDM	MIAS	DDSM
Projections	CC and MLO	MLO	CC and MLO
Spatial resolution	94 μm/pixel	50 μm/pixel	43.5, 50, 42 μm/pixel
Gray-level quantization	12 bits	8 bits	12, 16 bits
Dimension	2294 × 1914 pixels	1024 × 1024 pixels	variable

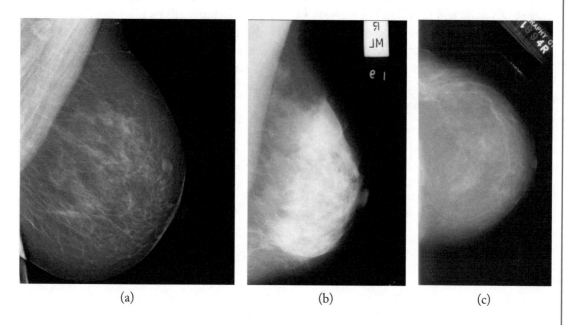

(a)　　　　　　　　　　(b)　　　　　　　　　　(c)

Figure 2.1: Examples of mammograms from (a) FFDM, (b) MIAS [141], and (c) DDSM [61] databases.

by means of a graphical user interface by Dr. Fabio F. Mangieri of the Diagnostic Radiology Unit, San Paolo Hospital of Bari, Italy, and by Dr. Maria Luisa Pepe of the Integrated Operational Unit of Diagnostic Imaging, Presidio Ospedaliero Occidentale, Castellaneta-Massafra-Mottola, ASL, Taranto, Italy. Additional ground-truth contours of images from the MIAS database were provided by Dr. Ricardo J. Ferrari from Departamento de Computação, Universidade Federal de São Carlos, São Paulo, Brasil, and his colleagues [50]. All of the annotated regions or landmarks together with the corresponding indication of presence of masses, regions of architectural distortion, or bilateral asymmetry were used as ground truth for detection studies. The results of biopsy were used as the ground truth for the diagnostic classification.

The BI-RADS classification of the MIAS images, used to evaluate the dependence of the methods' performance on mammographic density, were provided by J. Pont and E. Pérez of the Department of Radiology of the Girona University Hospital "Dr. Josep Trueta" (Spain) and E.R.E. Denton of the Department of Breast Imaging of the Norwich and Norfolk University Hospital (UK).

2.3   REMARKS

This chapter presented details regarding the three databases of mammograms used for validation of results and provided information on radiologists' annotations used to determine the performance levels of the various algorithms. The databases used include two publicly available databases of digitized SFMs and a private database of FFDMs. Additional details on the experimental setup for the acquisition of the various sets of images and of the corresponding ground-truth annotations are provided in the subsequent chapters, together with the description of the designed methods.

CHAPTER 3

Multidirectional Gabor Filtering

3.1 EXTRACTION OF DIRECTIONAL COMPONENTS

Directional filtering of images is a means to extract oriented patterns over specific directions of analysis. This is important especially when the structures or the objects to be analyzed possess an intrinsic directionality conveying information of interest for their characterization. The oriented patterns of the mammographic appearance of the breast parenchyma and the description of how such components are placed in the 2D space within the surrounding anatomical structures, in particular, play a key role in understanding and interpreting mammograms. Therefore, in addition to the gray-level image representation of the breast, the magnitude and phase images obtained by multidirectional Gabor filtering were prepared with the following purposes.

Edge detection: binarization of the Gabor magnitude response exploiting the high response values of pixels along the breast skin-line; more details are provided in Chapter 4.

Conditions on false edge points: suppression of structures of the breast subject to certain orientation by means of the phase response of the Gabor filters; more details are given in Chapters 4 and 7.

Enhancement of directional components: analysis of the Gabor magnitude response via the extraction of relevant features for quantification of directionality, as described in Chapters 5, 6, and 7.

More details on Gabor filters and multidirectional analysis are provided in the following section.

3.2 THE FAMILY OF GABOR FILTERS

In the 2D space, Gabor filters are complex, sinusoidal plane waves within a 2D Gaussian envelope, defined as

$$g(x, y) = \frac{1}{2\pi\sigma_x\sigma_y} \exp\left[-\frac{1}{2}\left(\frac{x^2}{\sigma_x^2} + \frac{y^2}{\sigma_y^2}\right)\right] \exp\left(2\pi j f x\right), \tag{3.1}$$

where σ_x and σ_y are the standard deviation values of the Gaussian function along the x and y directions, respectively, and f is the spatial frequency of the sinusoid. Gabor filters oriented at different angles can be obtained via a rigid rotation of the x–y coordinate system. Then, the frequency domain and the range of orientations of interest are covered by filtering the images with a set of Gabor filters with preferred spatial frequencies and orientations.

3.2.1   THE REAL GABOR FILTER

When compared with other types of filters, including Gaussian steerable filters and the complex Gabor filters themselves, the real Gabor filters yield the best performance in terms of capability to detect the presence of oriented features as well as in terms of accuracy in the estimation of the angles of orientation [4]. For this reason, the mathematical description of the methods for the extraction of oriented features and related applications will be presented with reference to the real Gabor filters, which also constitute some of algorithms designed in this work.

The real Gabor kernel, $g_r(x, y)$, oriented at $-\pi/2$ radians, corresponds to the real component of the function in Eq. (3.1), which is

$$g_r(x, y) = \frac{1}{2\pi\sigma_x\sigma_y} \exp\left[-\frac{1}{2}\left(\frac{x^2}{\sigma_x^2} + \frac{y^2}{\sigma_y^2}\right)\right] \cos(2\pi f x). \tag{3.2}$$

The Fourier transform of the Gabor filter determines its frequency domain representation and specifies the modulation of each frequency component of the input image. As $g_r(x, y)$ is real and even, the corresponding Fourier transform will be also real-valued as

$$G_r(u, v) = \exp\left\{-\frac{1}{2}\left[\frac{(u - u_0)^2}{\sigma_u^2} + \frac{v^2}{\sigma_v^2}\right]\right\} \exp\left\{-\frac{1}{2}\left[\frac{(u + u_0)^2}{\sigma_u^2} + \frac{v^2}{\sigma_v^2}\right]\right\}, \tag{3.3}$$

where $\sigma_u = 1/(2\pi\sigma_x)$ and $\sigma_v = 1/(2\pi\sigma_y)$. The values of σ_x and σ_y can be linked by the elongation parameter l, which determines the extent of the filter in the y direction as compared to the extent of the filter in the x direction. The scale of the filter can be determined by means of the thickness parameter, τ, that corresponds to the width of the filter at half maximum of the Gaussian term [4]. The stated parameters can be derived from the following design rules:

$$\begin{aligned} \sigma_y &= l\sigma_x \\ \sigma_x &= \tau/2\sqrt{2\ln 2} \\ f &= 1/\tau. \end{aligned} \tag{3.4}$$

The real Gabor kernel oriented at $-\pi/2$ radians, with $\tau = 6$ pixels and $l = 4$, and its Fourier transform are illustrated in Figs. 3.1a and b.

3.2.2   MULTIDIRECTIONAL FILTERING

The multidirectional approach to analysis of images consists of the use of a bank of Gabor filters oriented at different angles in the space domain. Specific directions of analysis can be explored

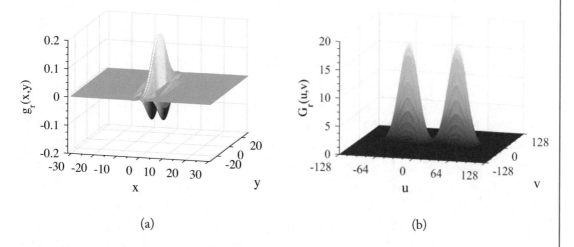

(a)   (b)

Figure 3.1: (a) Real Gabor kernel oriented at $-\pi/2$ radians, with $\tau = 6$ pixels and $l = 4$ and (b) the corresponding frequency representation.

by means of the coordinate transformation

$$\begin{bmatrix} x_k \\ y_k \end{bmatrix} = \begin{bmatrix} \cos(\alpha_k) & \sin(\alpha_k) \\ \sin(\alpha_k) & \cos(\alpha_k) \end{bmatrix} \begin{bmatrix} x \\ y \end{bmatrix}, \qquad (3.5)$$

that gives the set of coordinates (x_k, y_k) rotated by the angle α_k in radians. Let $I(x, y)$ be the image being processed. The corresponding filter response, $W_k(x, y)$, is obtained by convolving $I(x, y)$ with the kernel $g_k = g_r(x_k, y_k)$. Examples of four Gabor kernels oriented at different angles are shown in Fig. 3.2. If a bank of filters is generated for angles $\alpha_k = \pi k / K$, $k = 0, 1, \ldots, K$, where K is the number of equally spaced filters over the half-closed angular interval $(-\pi/2, \pi/2]$ radians, then the magnitude response, $W(x, y)$, and the phase response, $\Phi(x, y)$, are generated by assigning at each pixel the maximum value obtained among the filter responses and the corresponding angle, respectively [4]:

$$W(x, y) = \max_{k} \left[W_k(x, y) \right], \qquad (3.6)$$

$$\Phi(x, y) = -\frac{\pi}{2} + \arg\max_{k} \left[W_k(x, y) \right]. \qquad (3.7)$$

3.2.3 CHOICE OF FILTER PARAMETERS

The design parameters, τ and l, determine the size of the filter, which was set to $6\sigma_x + 1$ and, more importantly, the size of the oriented structures that the filter is capable of capturing. To

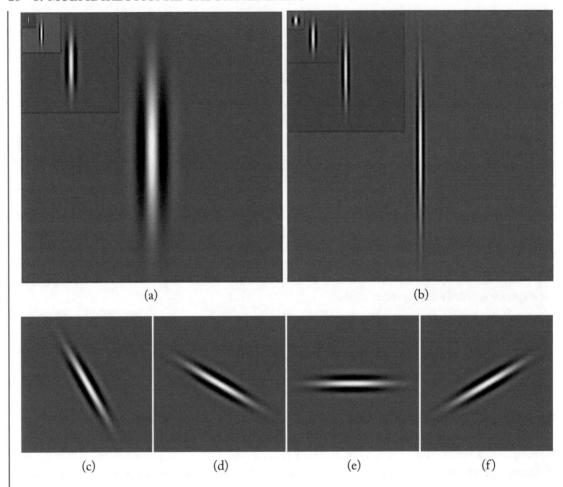

Figure 3.2: Examples of Gabor kernels in the spatial domain. (a,b) Kernels oriented at $-\pi/2$ radians for (a) $l = 4$ and increasing values of thickness, i.e., $\tau = 2, 6, 15$, and 40 pixels, and (b) $\tau = 10$ and increasing values of elongation, i.e., $l = 1, 3, 7$, and 16. (c,d,e,f) Kernels with $l = 4$ and $\tau = 20$ pixels oriented at different angles: (c) $\alpha = -\pi/3$, (d) $\alpha = -\pi/6$, (e) $\alpha = 0$, (f) $\alpha = \pi/6$ radians.

understand the variations induced on the filter by the parameters, the Gabor kernel oriented at $-\pi/2$ radians with $l = 4$ is shown in Fig. 3.2a for various values of τ, i.e., $\tau = 2, 6, 15$, and 40 pixels. If the thickness of the filter is fixed at $\tau = 10$ pixels, variations in the elongation produce the results shown in Fig. 3.2b for values $l = 1, 3, 7$, and 16. The effect of increasing values of the corresponding design parameter can be observed starting from the top left corner. If $K = 18$ is chosen, we have 18 kernels for the bank of filters and the absolute angular error between the angle of a given oriented pattern and the closest kernel angle in the filter bank will eventually span the range $[0, 5°)$.

As an example, directional analysis was performed on an image of an aquarium plant with a bank of 180 Gabor filters equally spaced over the range $(-\pi/2, \pi/2]$ radians. The original image of size 1000×1000 pixels is shown in Fig. 3.3a. The Gabor parameters used for the analysis were $l = 5$ and $\tau = 12$ pixels. The magnitude response image and the orientation field are shown in Figs. 3.3b and c, respectively, in which the various oriented leaves and corresponding orientation are emphazised. The response images for individual filters oriented at $-\pi/6$, 0, and $\pi/6$ radians are illustrated in Figs. 3.3d–f.

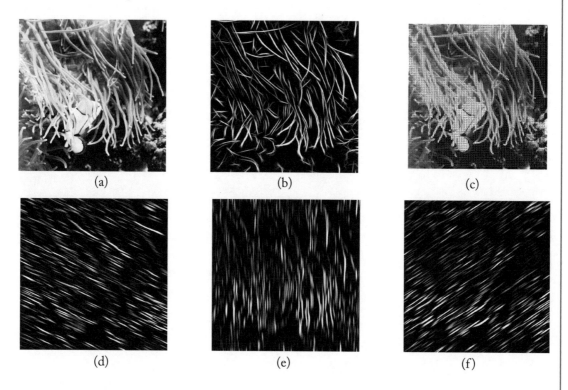

Figure 3.3: Directional analysis of an image of an aquarium plant. (a) Original image of size 1000×1000 pixels. (b) Magnitude response image. (c) Orientation field superimposed on the original image. Magnitude response images for kernel oriented at (d) $-\pi/6$, (e) 0, and (f) $\pi/6$ radians.

3.3   REMARKS

In this chapter, methods for the detection and analysis of oriented patterns using Gabor filters as originally developed by Ayres and Rangayyan [4] were discussed. The results of filtering using a bank of Gabor kernels are combined to form two images: the magnitude image, which indicates the intensity of the oriented feature at each pixel, and the orientation field, which indicates the orientation of the feature at each pixel. The real Gabor filter is recommended when the images under investigation possess intrinsic directionality and when high detection performance and angular accuracy are required, as in the case of mammographic images. In the present work, Gabor filters are used for the tasks of detecting the breast skin-line, for the characterization of mammographic lesions, and for detecting bilateral asymmetry, as described in Chapters 4, 5, 6, and 7.

CHAPTER 4

Landmarking Algorithms

4.1 LANDMARKING OF BIOMEDICAL IMAGES

Before being analyzed by a CAD system, biomedical images need to be processed in a preliminary stage in which the relevant anatomical regions or structures are segmented from the remaining portions of the images.

Effective landmarking is particularly important for analyzing mammographic images. For the CAD system for detection of bilateral asymmetry (described in Chapter 5), landmarking is used to derive anatomical correspondence between the left and right mammographic regions to be compared. For the integrated CADe/CADx system (described in Chapter 7), landmarking procedures are needed as a preliminary step to confine the region where abnormalities may be localized; they also help in reducing the FP outcomes due to the presence of intricate anatomic structures or bright areas that may affect the performance of the procedure for detection of masses and regions of architectural distortion. In mammograms, the anatomical structures to be detected are the pectoral muscle (only for MLO views), nipple, and breast skin-line.

The presence of the pectoral muscle can strongly influence the performance of a procedure for identification of dense tissue, such as fibroglandular tissue or bilateral asymmetry [17, 94]. Another reason to identify the pectoral muscle is that the orientation of the muscle is a reference for image registration, content-based image retrieval, and bilateral comparison.

A modified version of the method proposed by Ferrari et al. [49] for automatic detection of the pectoral muscle is described in the following sections. Two recently developed algorithms are presented that estimate the position of the nipple and the breast skin-line. The methods were tested on a diverse set of mammograms, including both SFMs and FFDMs. Before giving the details of the approaches, previous work dealing with detection of the pectoral muscle, detection of the nipple, and automatic extraction of the breast skin-line in mammograms are reviewed.

4.2 STATE OF THE ART

4.2.1 PREVIOUS WORK ON DETECTION OF THE PECTORAL MUSCLE

Only a few works have been presented in the literature that address the detection of the pectoral muscle. One of the first approaches was introduced by Karssemeijer [68]. The method is based on the Hough transform and on the use of a set of threshold values to detect a linear approximation of the pectoral muscle. The procedure was subsequently modified by Ferrari et al. [47] who reported average values of FP and FN rates of 1.98% and 25.19%, respectively, on a set of 84 mammograms from the MIAS database. Aylward et al. [3] used a method based on a gradient magnitude ridge traversal algorithm and on a voting scheme procedure.

To overcome the limitation of hypothesizing a straight line for the representation of the pectoral muscle, especially when the curvilinear nature of the pectoral muscle is evident on the mammogram, starting from the approach proposed by Manjunath and Ma [89], the group of Ferrari and colleagues [49] proposed the use of Gabor wavelets to extract the curvilinear profile of the pectoral muscle edge and achieved, on average, 0.58% and 5.77% FP and FN rates, respectively.

4.2.2   PREVIOUS WORK ON DETECTION OF THE NIPPLE

There are several methods proposed in the literature to identify the nipple in mammograms. One of the first attempts was made by Yin et al. [164], who considered the maximum of the average gray-level profile computed along a selected portion of the anterior border of the breast in order to locate the nipple and perform automated alignment of digitized mammograms. They reported a mean error of 10 mm over a total of 80 images. Karnan and Thangavel[67] combined the results obtained by maximizing the height of the breast border and the second derivative across the median-top section of the breast. The mean error reported in their study was 13.5 mm, using 156 images. Chandrasekhar and Attikiouzel [30] used a similar approach to locate the nipple. Even though they reported an error of less than 1 mm in 96% of images, their method was tested on a small dataset consisting of 24 mammograms. They also mentioned that their method could fail when the nipple is noticeably recessed or when benign or malignant processes modify the intensity profile of the image.

Based on the convergence of the texture pattern toward the nipple, Zhou et al. [171] developed a method for texture orientation-field analysis to estimate the nipple location. They reported a mean error of 2.5 mm on 367 randomly selected digitized mammograms; however, 2.5% of the cases had errors larger than 50 mm. Following a similar approach, Kinoshita et al. [73] developed a method for automatic detection of the nipple via image processing in the Radon domain. Their method was tested on a private dataset containing 1080 digitized mammograms and provided an average error of 7.4 mm. Iglesias and Karssemeijer [63] proposed a multiatlas algorithm capable of finding the nipple with a low number of outliers (0.13% of cases had an error larger than 50 mm in a set of 2340 SFM images); they obtained a mean error of 12 mm. However, images that did not contain the nipple in their field of view were discarded during performance evaluation. The location of the nipple, as proposed by van Engeland et al. [152], can be estimated as the point on the skin contour with the largest distance to the chest or the pectoral muscle. The method was described to have an average error of 14 mm on digitized mammograms, not considering images where the nipple was not in the field of view. To our knowledge, all of the available methods for the detection of the nipple have been developed only for digitized SFM images. However, FFDM is gradually replacing SFM, and the increasing need of comparison between mammograms will require robust procedures that are able to process mammograms acquired with different technologies.

4.2.3 PREVIOUS WORK ON SEGMENTATION OF THE BREAST REGION

Several studies have addressed the problem of automatically detecting the breast boundary in mammograms. The early reported attempts were based on histogram thresholding and morphological filtering operations [36, 164]. Despite the advantages of simplicity and convenience of these approaches, they were prone to errors due to a critical dependence of the result on the threshold selection process. The presence of background noise and artifacts also reduced the chances of obtaining an accurate estimate of the breast skin-line. Therefore, subsequent work considered further refinement procedures, applied to an approximate breast boundary extracted in a preliminary step, in order to improve the measures of acceptability of the results. Bick et al. [10] proposed a modified method for global histogram analysis followed by a combination of a gray-value range operator and region growing. Méndez et al. [96] divided the mammogram into three regions and used a tracking algorithm, consisting of gradient-based conditions applied to nine neighboring pixels at each point, to detect the border.

The first quantitative evaluation of segmentation results was provided by Ojala et al. [104]. In their work, a preliminary segmentation was performed using an adaptive histogram thresholding method and morphological filtering; the final breast boundary was then extracted by means of three different smoothing procedures: filtering in the Fourier domain, snakes, and B-splines. The best reported result, in terms of the mean error between the manually delineated and the automatically computed breast boundary, was achieved using the spline technique. The reported average error was 2.2 mm over 20 test images.

Ferrari et al. [50] used an active deformable contour model especially designed to be adaptive. The model was initialized by an approximate breast boundary obtained using the Lloyd-Max quantization method, morphological opening operators, and the chain-code method. The influence of the initial contour on the convergence of the algorithm was minimized by including a balloon force in the energy formulation. The reported average FPR and FNR were 0.41% and 0.58%, respectively, on a total of 84 mammograms.

A contour growing technique was proposed by Martí et al. [91] using scale-space edge detection with attraction and regularization terms. Their method was tested on 65 images from the MIAS database [141] and provided mean correctness and completeness values of 0.97 ± 0.06 and 0.96 ± 0.06, respectively. In addition, 24 images from the DDSM [61] were tested and average values of 0.95 ± 0.06 and 0.97 ± 0.01 were reported for correctness and completeness, respectively.

Raba et al. [118] used a region growing approach and tested their method over the whole MIAS database [141]. They obtained a "near accurate" result for 98% of the mammograms.

Sun et al. [142] introduced the concept of dependency in distance from the stroma edge to the skin-line. The stroma edge was initially extracted by Otsu's thresholding method [107].

A combination of adaptive thresholding and the greedy range selection method was used to estimate only the portion of the skin-line around the central nipple area. Due to the presence of background noise and artifacts, the missing upper and lower portions were extrapolated from the stroma edge using average Euclidean distances. They reported an average polyline distance measure (PDM) error of 3.28 pixels (0.66 mm) for 82 mammograms from the MIAS database [141]. The same dataset was used to perform a quantitative comparison with the results obtained by Ferrari et al. [50], which yielded a mean PDM of 4.92 pixels (0.98 mm).

Yapa and Harada [163] developed a fast marching algorithm to track the evolution of the breast skin-interface by means of a partial-differential equation. They reported 99.1% accuracy in the segmentation results using 100 mammograms from the MIAS database.

Karnan et al. [67] applied a genetic algorithm to binary images of mammograms obtained by histogram thresholding. They evaluated the accuracy of the breast boundaries detected with their method using Pratt's figure of merit, which is an approximate indicator of edge quality, and obtained an average measure of 0.93 for 114 images.

Padayachee et al. [108] proposed a new method for selecting the optimal gray-level threshold by analyzing the areas enclosed by isointensity contours. The algorithm was tested on 25 mammograms and yelded an average root-mean-squared error of 4.08 mm, using a pre-processing step based on the Lorentzian kernel.

In the work of Wu et al. [162], an initial breast boundary, obtained by dynamic thresholding, was refined by using gradient information from horizontal and vertical Sobel filtering. The accuracy of the breast boundary detection algorithm was evaluated on a total of 716 SFMs, by comparison with the reference boundaries drawn by radiologists. The reported Hausdorff distance was less than 4.8 mm for 94% of the images.

Silva et al. [135] used a dynamic programming algorithm by means of the definition of a cost function and reported an average PDM of 0.41 mm. Kus and Karagoz [79] developed a system to automatically detect the breast boundary by applying a multidirectional scanning window and constraints on the intensity and gradient pixel values. They reported average Hausdorff distance and PDM of 2.19 mm and 0.35 mm, respectively. The results reported by Silva et al. [135] and Kus and Karagoz [79] were obtained on the same set of 82 images used by Ferrari et al. [50] and Sun et al. [142]. Additional work on detection of the breast skin-line includes the use of region growing [168] and active contour [93, 95, 148] algorithms; however, the related reports presented only qualitative evaluation of the results.

It is important to note that in all of the works mentioned above, the performance of the algorithms was not evaluated on an independent test set of mammograms. This could lead to results which are dependent on the parameters chosen for the particular set of images. Another pitfall identified in this book concerns the lack of testing procedures for detection of the breast boundary on both SFMs and FFDMs. In fact, while digitized films are still encountered in clinical practice, FFDMs are gradually replacing SFMs, and the validation of a unique and robust procedure on both imaging modalities will benefit automated analysis of mammograms.

4.3   DETECTION OF THE PECTORAL MUSCLE

4.3.1   OVERVIEW OF THE METHODS

A simple method for the extraction of the pectoral muscle is based on the Hough transform [49]. It derives a linear representation of the muscle, thus introducing a simplification of the real anatomy as represented on the image. The method has the advantage of being fast and simple.

A more sophisticated strategy, which overcomes the straight-line representation of the previous approach, is based on Gabor wavelets and, in particular, on the procedure developed by Ferrari et al. [49]. The blocks in common with the existing approach used in this book deal with image decomposition using Gabor wavelets and computation of the magnitude and phase images by vector summation. By performing a different selection of angles for the Gabor wavelets, we could eliminate the phase propagation step and the correction of disjoint boundary segments. The reduction of the false lines detected at the first step was performed by considering five different conditions, later denoted as C_1, C_2, C_3, C_4, and C_5, that consider the orientation, the position, and the length of each line and, in addition, the local mean intensity values along and around each line detected to select the most probable muscle profile [95]. More details on both approaches are provided in the following sections.

4.3.2   DATASET AND EXPERIMENTAL SETUP

A set of cases was randomly selected from the three databases described in Chapter 2: 141 pairs of MLO mammograms from the DDSM [61], 45 pairs of mammograms from the MIAS database [141], and 20 pairs of mammograms from the FFDM database. All the images were preliminary downsampled to a spatial resolution of 300 μm using a bicubic interpolation method, and flipped as required so that the pectoral muscle was always in the left-upper portion of the MLO images.

A preliminary evaluation of the method was performed by acquiring the opinion of an expert radiologist. To facilitate the acquisition even by radiologists from different countries, a web-based platform has been developed, written in *Perl-Language*, that allows a simple evaluation of the results produced by the method by any radiologist. The radiologist can assign a score to the result of the pectoral muscle boundary extraction. The available scores, selected using a drop-down box, are *Excellent*, *Good*, *Average*, *Poor*, and *Complete Failure*.

4.3.3   METHODS

Detection of the Pectoral Muscle Line

The method for the extraction of the pectoral muscle based on the Hough transform is summarized as follows.

Step 1: A Gaussian filter with $\sigma_x = \sigma_y = 4$ pixels is used to smooth the images in order to remove the high-frequency noise in the image.

Step 2: Mammograms of right breasts are flipped (mirrored).

Step 3: The Hough transform is then applied to the Sobel gradient of the images.

Step 4: Only values of θ in the range [30 ÷ 80] are considered in the analysis.

Step 5: The parameters ρ and θ with the maximum value are taken to represent the pectoral muscle line.

The simplest case of the Hough transform is the linear transform for detecting straight lines. In the image space, a straight line can be specified in terms of its orientation, with respect to the x axis, and its distance from the origin. In this form of parametrization, any straight line is bounded in angular orientation by the interval $[0, \pi]$ radians and bounded by the Euclidean distance to the farthest point of the image from the center of the image, as shown in Fig. 4.1. The equation for an arbitrary straight line segment in the image plane is given by

$$\rho = x\cos(\theta) + y\sin(\theta). \tag{4.1}$$

For a specific point in the image domain, (x_i, y_i) we obtain a sinusoidal curve in the Hough domain (ρ, θ). For each point (x_{i_0}, y_{i_0}) lying on a straight line with $\rho = \rho_0$ and $\theta = \theta_0$ in the image domain, the related sinusoidal curve in the Hough domain is specified by

$$\rho_0 = x_{i_0}\cos(\theta_0) + y_{i_0}\sin(\theta_0). \tag{4.2}$$

Through Eq. (4.2), it is evident that, for each point in the image domain, the Hough transform performs a one to many mapping, resulting in a modulated sum of sinusoids in the Hough domain.

The Hough transform is often referred to as a voting procedure, where each point in the image casts votes for all parameter combinations that could have produced the point. All of the sinusoids resulting from the mapping of the points in a straight line in the image domain have a common point of intersection at (ρ_0, θ_0) in the Hough domain.

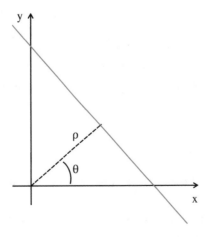

Figure 4.1: Parametric representation of a straight line.

The following properties of the above representation are to be noted [49].

- A point in the image space (x, y) corresponds to a sinusoidal curve in the (ρ, θ) Hough parameter space.

- A point in the (ρ, θ) space corresponds to a straight line in the (x, y) space.

- Points lying on the same straight line in the (x, y) space correspond to curves through a common point in the parameter space.

- Points lying on the same curve in the parameter space correspond to lines through a common point in the (x, y) space.

Linear segments in the spatial domain correspond to large-valued points in the Hough domain (see Fig. 4.2). Thus, the problem of determining the directional content of an image becomes a problem of peak detection in the Hough parameter space.

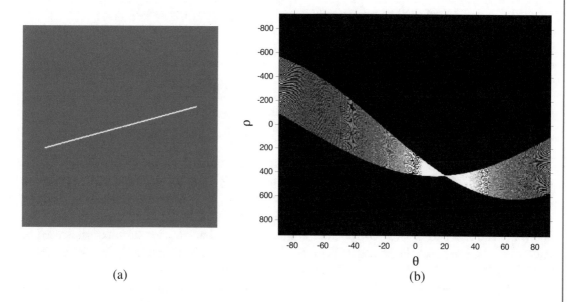

(a)   (b)

Figure 4.2: (a) Straight line with $\rho = 400$, $\theta = 20°$. (b) Hough space for the line in (a).

For detecting the pectoral muscle line, the gradient magnitude $M_g(x, y)$ and orientation $\theta_g(x, y)$ of the image $f(x, y)$ are computed by applying a 3×3 Sobel operator [68]. The gradient orientation, $\theta_g(x, y)$, is used in the following condition: $|\theta_g(x, y) - \theta| < \delta\theta$. The histogram of non-zero gradient magnitude values, h_g, is used to determine an increment value giving a larger weight in the Hough space to pixels with a strong gradient in the image. The weight function,

$w(g)$, that is used is defined by

$$w(g) = \frac{1}{A} \int_0^n h_g(n)\, dn \tag{4.3}$$

with A equal to the area under h_g. In addition, a rectangular area in the Hough space including values with $\rho < 0$ and $0.7\,\pi < \theta < 0.98\,\pi$ is set to zero. Also, the contribution of pixels to a given peak is limited by some fixed threshold, l_{min}, on the length of the projected line $l(\rho, \theta)$ [68]. This ensures that the right peak in the Hough space is selected and the pectoral muscle line can be found.

Detection of the Pectoral Muscle Profile

As described with more detail in Chapter 3, the 2D Gabor function is a Gaussian function modulated by a complex sinusoid, specified by its frequency W and the two standard deviation values σ_x and σ_y of the Gaussian envelope. The Gabor function is given by

$$\psi(x, y) = \frac{1}{2\pi\sigma_x\sigma_y} \exp -\frac{1}{2}\left(\frac{x^2}{\sigma_x^2} + \frac{y^2}{\sigma_y^2}\right) \exp(j\,2\pi W x)$$

to which corresponds, in the frequency domain, the filter

$$\Psi(u, v) = \begin{cases} \frac{1}{2\pi\sigma_u\sigma_v} \exp\left(-\frac{1}{2}\left(\frac{(u-W)^2}{\sigma_u^2} + \frac{v^2}{\sigma_v^2}\right)\right), \\ \qquad \text{for } (u, v) \neq (0, 0), \\ 0 \quad \text{for } (u, v) = (0, 0), \end{cases}$$

where $\sigma_u = 1/(2\pi\sigma_x)$ and $\sigma_v = 1/(2\pi\sigma_y)$. Using the same relations illustrated by Ferrari et al. [49] for all the parameters describing the wavelet family, we used orientations at 5, 10, 15, 20, 25, 30, 35, and 40° and a number of scales $s = 3$, keeping in mind that perfect reconstruction is not required in this kind of analysis. The magnitude and the phase of the response of the real part of each Gabor filter is then computed and vector summation is used to sum up all the responses. Further details on the filters used can be found in the work by Ferrari et al. [49]. The visual result of this preliminary step is given in Fig. 4.3b.

The second step is a preliminary search for boundaries represented by points having opposite phase orientation. Denote the resulting image with M. After this step, a binarization with adaptive threshold equal to $0.01 \times \max(M)$ is performed to obtain possible candidates for pectoral muscle contours. An example is shown in Fig. 4.3c.

The third step deals with the removal of false lines by the application of the logical conditions $C_1 - C_5$. Consider Fig. 4.4. For each distinct object O_k, $k = 1, \ldots, N_{objects}$ shown in Fig. 4.3c and computed using an 8-connection rule, the following logical conditions were applied so as to maintain the corresponding object only if all the conditions are simultaneously satisfied.

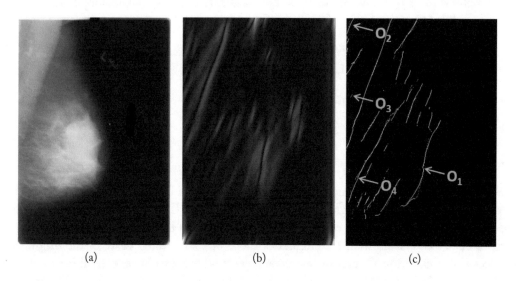

Figure 4.3: (a) Original image. (b) Magnitude of Gabor wavelets after vector summation. (c) Candidates for pectoral muscle profile. The blue superimposed arrows indicate false lines to be removed.

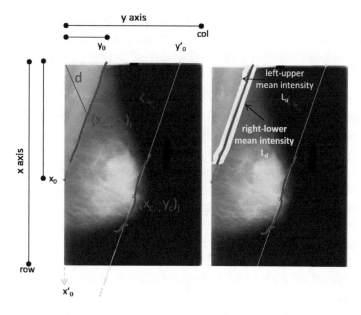

Figure 4.4: A visual representation of conditions C_1–C_5 for the reduction of false pectoral muscle boundary lines. Reproduced with permission from Mencattini et al. [95] © Springer.

1. Condition C_1: $y_0 < col$, where y_0 is the horizontal intercept of each object O_k extended toward the y axis and col is the width of the image.

2. Condition C_2: $x_0 < row$, where x_0 is the vertical intercept of each object O_k extended toward the x axis and row is the height of the image.

3. Condition C_3: $d > 0.9 \times \min(x_0, y_0)$.

4. Condition C_4: $number\ of\ pixels \in O_k > 10$.

5. Condition C_5: $L_u > L_d$ (see Fig. 4.4).

Where $(x_c, y_c)_k$ are the coordinates of the centroid of each object O_k and d represents the distance of the point $(x_c, y_c)_k$ from the origin of the axes. Conditions C_1 and C_2 eliminate those objects whose orientation, extended toward the axes, produces x_0 or y_0 outside the image (such as object O_1). Condition C_3 eliminates those objects located close to the origin (such as object O_2). Condition C_4 eliminates those objects whose area is very small (as object O_3). Condition C_5 eliminates those objects that do not appear to be the boundary between a light region (in the left-upper part) and a dark region (in the right-lower part) (as object O_4), since L_u and L_d represent the mean intensity value evaluated in a strip of 5 pixels in the upper-left part of the candidate object and in the lower-right part, respectively (see Fig. 4.4).

The five conditions are combined using the logic *AND* operator. At the end of this procedure, more than one object is likely to be a candidate for the pectoral muscle boundary (see Fig. 4.4). The largest line is finally chosen as the pectoral muscle profile.

4.3.4 RESULTS AND DISCUSSION

An example of the extraction of the pectoral muscle line based on the Hough transform is shown in Fig. 4.5. The method assumes that the edge of the pectoral muscle is approximately a straight line oriented in a certain direction.

For the case of the method based on Gabor wavelets, some examples of the extraction of the pectoral muscle profile are shown in Fig. 4.6. The method is robust to the presence of pectoral muscle minoris and of confounding densities. The histogram shown in Fig. 4.7 reports the scores assigned by the radiologist. Here, we considered only the MLO images, thus getting a total of 250 images from the three databases. Figure 4.8 illustrates two examples of an *Excellent* and a *Complete Failure* assessment provided by the radiologist.

An average Hausdorff distance of 3.77 mm was obtained by comparing the results of pectoral muscle detection with Gabor wavelets to the hand-drawn pectoral muscle edges available for the 84 MIAS mammograms. Manually marked pectoral muscle edges were not available for the DDSM and FFDM images and the related accuracy of the pectoral muscle detection could not be estimated. However, visual inspection of the segmentation results obtained for the other sets of images confirmed the effectiveness of the approach.

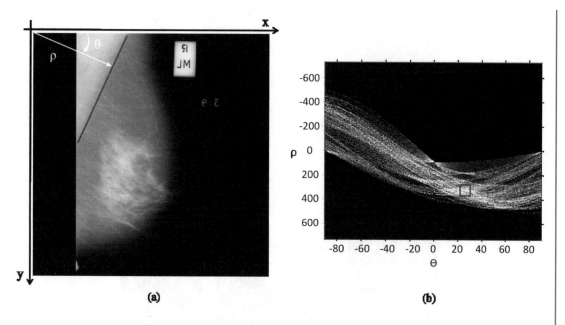

Figure 4.5: (a) Pectoral muscle line detected using the Hough transform. The coordinate system used to compute the Hough transform is also shown. (b) Hough transform of the Sobel gradient of the image. The red box is centered on the peak corresponding to the red line in (a).

4.4   DETECTION OF THE NIPPLE

Automatic identification of the nipple in mammograms is another fundamental step in the development of various algorithms that constitute CAD systems for mammography, in particular, the most recent and advanced computerized techniques to achieve high performance in terms of detection and diagnosis of subtle signs of breast cancer [72, 111, 124, 125, 145, 149, 150, 152, 165, 166, 169]. Such advanced approaches consist of multiple-view analysis of mammograms and can require point-based alignment and registration procedures in order to optimize comparison between images. Possible applications of methods for detection of the nipple are:

1. matching candidates in different mammographic projections of the same subject [111, 152, 165, 169];

2. simultaneous analysis of the current and prior mammograms of the same subject to recognize changes in the breast [72, 124, 125, 149, 150, 166];

3. bilateral comparison of the left and right breasts to analyze asymmetry and relative abnormalities as signs of lower-stage breast cancer [124, 150];

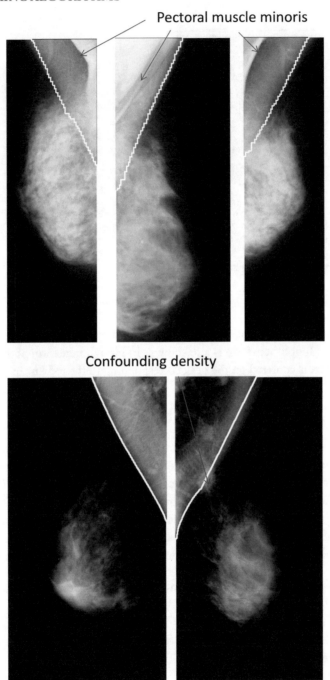

Figure 4.6: Examples of extraction of the pectoral muscle profile with Gabor wavelets.

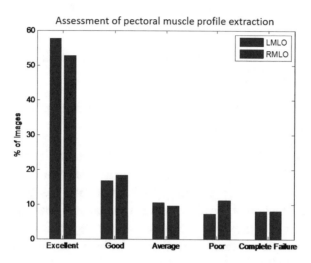

Figure 4.7: Qualitative evaluation of the results of pectoral muscle extraction with Gabor wavelets by an expert radiologist, performed on a total of 250 images. Reproduced with permission from Mencattini et al. [95] © Springer.

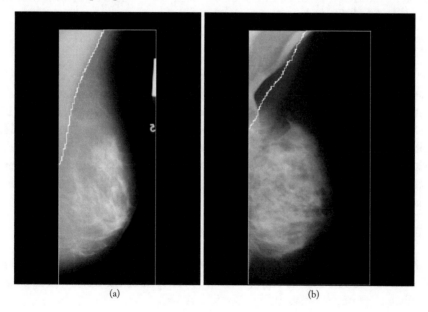

Figure 4.8: Examples of two pectoral muscle profiles obtained with Gabor wavelets and evaluated by the radiologist as: (a) *Excellent* and (b) *Complete Failure*.

4. masking procedures aimed at reducing FPRs and maintaining high levels of sensitivity via a systematic approach to the analysis of mammograms [145];

5. integration of lesion information acquired with different imaging modalities for improved diagnostic efficiency [166]; and

6. content-based image retrieval (CBIR) for quantitative description of mammograms and for assessing measures of similarity [72].

The accuracy in the detection of the nipple is crucial, as it is the only consistent and stable landmark on a mammogram. Developing robust methods for identification of the nipple can improve the potential of multiview analysis of mammograms in improving the chances of survival of cancer patients. However, detection of the nipple is still a difficult task owing to variations in image quality, nipple position, and appearance in projected mammographic images. Compression applied during the examination can cause distortion of the breast tissue, resulting in retraction of the nipple. Changes in positioning of the breast can lead to displacement of the nipple. Moreover, the use of different mammographic acquisition techniques, such as SFM, FFDM, and imaging devices that have different characteristics in terms of spatial resolution or contrast, can make the identification of the nipple difficult by using a single automatic procedure. Regardless, increasing use of multiview systems indicates the need for a single algorithm that works with images acquired with different imaging modalities.

4.4.1   OVERVIEW OF THE METHODS

We present a method for automatic detection of the nipple which is applicable to any kind of mammographic image. The procedure combines geometrical constraints with information on the gradient vector field (GVF). After automatic extraction of the breast region by identifying the breast skin profile, in CC and MLO views, as well as the pectoral muscle in MLO views, detection of the nipple is divided into two main steps: selection of a search region on the mammogram where the nipple is most likely to be located, and analysis of the second-order structure of the selected region by means of the eigenvalues of the Hessian matrix [18]. The following assumptions are made in this study.

1. The nipple is located near the breast skin line.

2. The nipple is located close to the farthest point on the skin line along a line that is perpendicular to the pectoral muscle edge (for MLO views) or to the chest wall approximated by the vertical edge of the breast image (for CC views).

3. The GVF locally converges at the center of the nipple.

4. The rate of convergence is approximately the same in every direction.

The first and second assumptions are used to extract a plausible nipple/retroareolar area (PNRA) on the mammogram by means of the breast skin line and the least-squares regression slope of the pectoral muscle edge (in MLO views). The morphological top-hat and Gaussian filters are applied in order to enhance the circular structure of the nipple. Based on the last two assumptions, the GVF of the PNRA is computed. Local shape-based constraints obtained via mean and Gaussian curvature measurements as well as the condition number of the Hessian matrix are used to select a cluster of pixels with certain specified properties in terms of local intensity variations. The centroid of the selected region is finally identified as the center of the nipple.

The flowchart in Fig. 4.9 gives an overview of the procedure outlined above. More details are provided in the following sections.

4.4.2   DATASET AND EXPERIMENTAL SETUP

For this study, a total of 566 mammographic images were randomly selected from multiple sources, including 194 FFDMs, 90 mammograms (MLO views) from the MIAS database, and 282 from the DDSM (142 CC views and 140 MLO views).

In order to process the mammograms in the same way, all of the images were downsampled to a spatial resolution of 300 μm/pixel using a bicubic interpolation method. The selected mammograms were annotated by a radiologist specialized in mammography (Dr. Maria Luisa Pepe), who provided the ground truth for the nipple position using a graphical user interface; in cases where the nipple was nearly invisible, the radiologist provided an estimate of the location of the nipple.

4.4.3   METHODS

Extraction of the PNRA

Even if the exact location of the nipple is not yet known, it is still possible to define a region of the breast where the nipple is most likely to appear on the mammogram. Defining a small search region on the mammogram can reduce the chance of false detection of the nipple caused by noise, artifacts, or by the presence of benign or malignant processes. This search region is referred to as the PNRA and it is defined so as to include the position of the nipple even in cases of retraction or displacement of breast structures. Our hypothesis is that the nipple is located close to the farthest point on the skin line along the line perpendicular to the pectoral muscle (in MLO views) or to the chest wall (in CC views), not far from the skin line. Hence, localization of the PNRA requires extraction of the breast skin contour as well as estimation the orientation of the pectoral muscle (for MLO views). An approximate breast boundary was extracted by histogram-thresholding and the morphological operation of closing. Only the largest area in the result was retained, thus eliminating labels and other unwanted structures in the background. A localized version of the active contour model without edges, proposed by Chan and Vese [29], was used to refine the rough boundary of the obtained binary mask, thus producing the breast skin contour. By knowing the orientation of the pectoral muscle (for MLO views) and the laterality of the

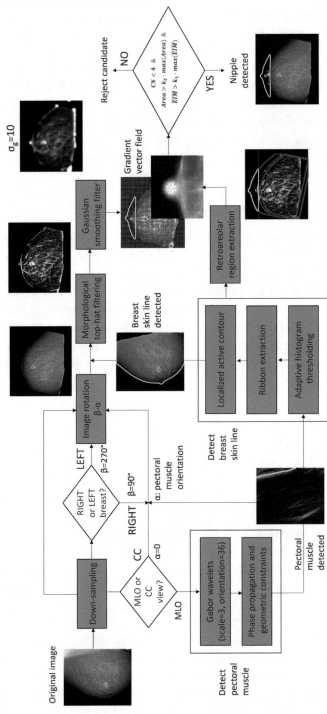

Figure 4.9: Flowchart of the procedure for estimation of the nipple, including the pre-processing steps for extraction of the pectoral muscle profile and the breast skin-line.

mammograms, images of right breasts were mirrored about the vertical axis. All of the images and their binary masks were then rotated so that the chest wall or the pectoral muscle was at the bottom of the image and parallel to the horizontal axis. This step resulted in a 90-degree rotation for CC views and a $(90 + \alpha)$-degree rotation for MLO views, where α is the angle between the vertical axis and the straight-line regression of the pectoral muscle. Figure 4.10a illustrates the result of extraction of the breast contour after the rotation step is applied. The area between the inner and the outer portions of the estimated breast skin line was selected by two 40-pixel (1.2 cm) vertical shifts applied to the breast contour (see Fig. 4.10b). The obtained region was further bounded by a horizontal line passing through the highest point of the inner contour, as shown in Fig. 4.10c.

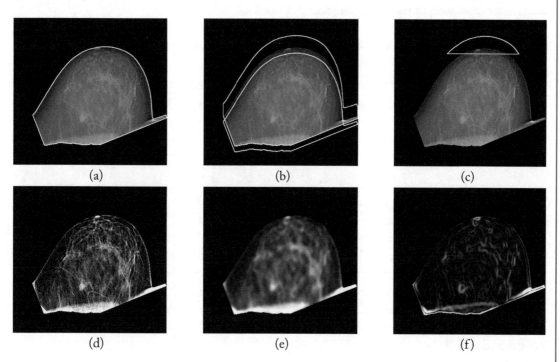

Figure 4.10: (a) Rotation of the image and the relative breast contour with reference to the pectoral muscle orientation. (b) Selection of the inner and the outer contours. (c) Extraction of the PNRA. (d) Result of top-hat filtering. (e) Result of Gaussian smoothing filtering ($\sigma_g = 10$ pixels). (f) Magnitude of the gradient.

Morphological Top-hat Filtering

The morphological white top-hat filter was applied to the rotated mammograms in order to enhance the circular structures present in the breast while darkening the rest of the breast tissue.

A disk-shaped structuring element with a radius of 30 pixels (9 mm) was used for this purpose (see an example in Fig. 4.10d).

Gaussian Smoothing Filtering

Images were filtered using a Gaussian filter with a standard deviation of 10 pixels (3 mm) to retain only low-frequency information (such as the core region of the nipple) and remove noise and fine structures (see Fig. 4.10e).

Analysis of the GVF

A common approach to analyze the local intensity behavior of a given image, I, is to consider the Taylor expansion in the neighborhood of a pixel $\mathbf{x_0}$:

$$I(\mathbf{x_0} + \Delta\mathbf{x}) \approx I(\mathbf{x_0}) + \Delta\mathbf{x}^T \nabla I(\mathbf{x_0}) + \frac{1}{2}\Delta\mathbf{x}^T \mathbf{H}(\mathbf{x_0})\Delta\mathbf{x}, \tag{4.4}$$

where $\Delta\mathbf{x}$ is a small increment about the pixel $\mathbf{x_0}$, ∇I is the gradient vector, $\nabla I = \left(\frac{\partial I}{\partial x_1}, \frac{\partial I}{\partial x_2}\right)$ and $\mathbf{H}$ is the Hessian matrix embedding the second partial derivatives of I as

$$\mathbf{H} = \begin{bmatrix} \frac{\partial^2 I}{\partial x_1^2} & \frac{\partial^2 I}{\partial x_1 \partial x_2} \\ \frac{\partial^2 I}{\partial x_1 \partial x_2} & \frac{\partial^2 I}{\partial x_2^2} \end{bmatrix}. \tag{4.5}$$

The second-order local structure of the image can be evaluated by analysis of the eigenvalues of $\mathbf{H}$. The method described in this study is inspired by the work by Wei et al. [160] and Mencattini and Salmeri [93], who used Hessian features to detect mass candidates. In this work, a novel approach was developed by considering measurements of the mean curvature (H) and the Gaussian curvature (K), which are taken as the sum and the product of the two eigenvalues (k_1 and k_2) of the Hessian matrix, respectively, computed locally for the PNRA. Figure 4.11a illustrates a scaled version of the PNRA and the corresponding GVF that locally converges at the nipple. Based on the signs of H and K, the local topographic structure of the PNRA can be determined as follows:

- if $K = 0$ and $H < 0$ the topographic structure is ridge-shaped,

- if $K = 0$ and $H > 0$ the topographic structure is valley-shaped,

- if $K = 0$ and $H = 0$ the topographic structure is planar,

- if $K > 0$ and $H < 0$ the topographic structure is ellipsoidal and peaked,

- if $K > 0$ and $H > 0$ the topographic structure is ellipsoidal and cupped,

- if $K < 0$ and $H < 0$ the topographic structure is a saddle, which is predominantly ridge-shaped, and

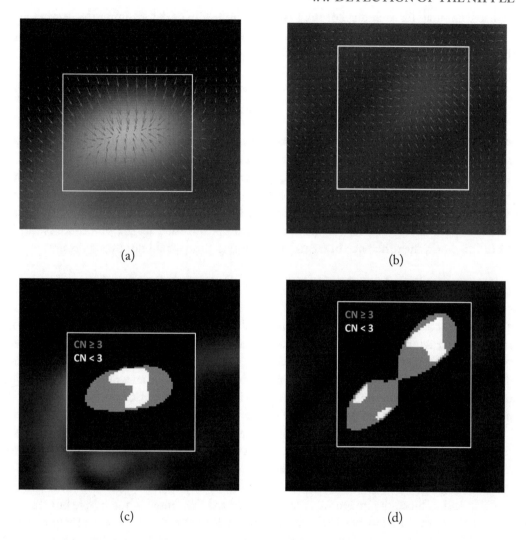

Figure 4.11: Zoomed view of the GVF superimposed: (a) on the structure of the nipple and (b) on a benign elongated structure in the mammogram. (c,d) Pixels satisfying the conditions $K > 0$ and $H < 0$ (white and gray) and $CN < 3$ (white).

- if $K < 0$ and $H > 0$ the topographic structure is a saddle, which is predominantly valley-shaped.

In order to characterize the topographic structure of the PNRA, the magnitude of the gradient was first computed from the filtered image (see Fig. 4.10f), and then measurements of H and K were derived for each pixel of the PNRA. Moreover, the singular values of the Hessian matrix were also computed for each pixel of the PNRA with the aim of determining the isotropicity of the level lines of I in the neighborhood of x_0, by means of the condition number (CN), which is the ratio of the larger to the smaller singular values of the Hessian matrix:

$$CN = \frac{\max(k_1, k_2)}{\min(k_1, k_2)}. \tag{4.6}$$

In particular, the condition $CN = 1$ corresponds to the center of a perfectly radially symmetric structure, whereas an elongated topographic structure has $CN \gg 1$. Note that even though applying the filters only to the PNRA is computationally less expensive, in this work, the filters were applied to the whole image and then the measurements were derived for the PNRA, hence avoiding the problems related to the definition of its border and the extent of its effects. In fact, since the images were previously downsampled and optimization of the procedure is not the focus of this work, the difference in terms of computing time would not be significant.

Local Shaped-based Constraints

Based on the last two assumptions given in Section 4.4.1, the nipple results in a locally ellipsoidal and peaked structure in which the local intensity of the gradient increases uniformly toward the center of the structure. Figure 4.11a illustrates a scaled version of the PNRA and the corresponding GVF that locally converges at the nipple. Note that the rate of convergence is almost uniform in every direction. In Fig. 4.11b, an example of the GVF for an elongated benign structure is given, in which the rate of convergence varies with the direction, decreasing along the direction of the maximum elongation of the structure.

These assumptions led to the definition of local shaped-based constraints to select a cluster of pixels with the specified properties. In particular, constraints were placed so that the candidates had $K > 0$ and $H < 0$, hence belonging to a locally ellipsoidal and peaked structure. All of the pixels satisfying these two conditions are colored in white and gray within the white frame in Figs. 4.11c and d. Since the image value increases toward the center of the nipple but the slope is approximately the same in every direction, only pixels with $CN < 3$ were considered as possible candidates. In this way, pixels belonging to elongated and peaked benign structures, such as the one shown in Figs. 4.11b and d, were mostly rejected, retaining most of the ellipsoidal structures characteristic of the nipple. The constrained method yielded a map of disjoint regions, whose pixels satisfied all of the previously stated conditions. Such selected regions are shown in Fig. 4.12a. In order to reject candidates with a small area, the areas of all of the selected regions were computed and only the top 50% in a ranked list in decreasing order of area were selected (see the selected areas in Fig. 4.12b). Among the remaining regions, the one with the maximum average Gaussian curvature was selected and its centroid was taken as the center of the nipple, as shown in Fig. 4.12c.

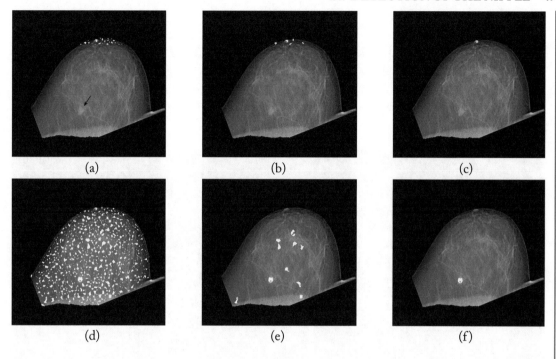

Figure 4.12: Maps of disjoint regions whose pixels satisfied the local shape-based conditions superimposed on the rotated version of mammogram with a malignant tumor (indicated by the arrow). Results obtained (a,b,c) with and (d,e,f) without defining the PNRA. (a,d) Areas satisfying conditions $K > 0$ and $H < 0$ and with $CN < 3$. (b,e) Candidates after rejection of small regions. (c,f) Final candidate. The result in (c) is the desired response.

4.4.4   RESULTS AND DISCUSSION

The algorithm was applied to three different types of mammograms, including SFM (mini-MIAS and DDSM) and FFDM images, taking into account the presence of noise, artefacts, and benign or malignant processes. As an example, results of the experiments performed on a mammogram in the presence of a malignant tumor are shown in Fig. 4.12 in order to understand the key role of the PNRA in reducing the chance of false detection of the nipple. The local shape-based conditions were applied to the whole breast region after filtering, and the obtained results (see Fig. 4.12d) are compared with the results obtained by defining the PNRA (see Fig. 4.12a). The results show that while the geometrical constraints introduced with the PNRA reduce the number of candidates and lead to the detection of the nipple (Fig. 4.12c), in the case without the definition of a selected search area, the final candidate corresponds to the tumor (Fig. 4.12f).

To evaluate the efficiency of the method, the Euclidean distance between the detected position and the center of the nipple as identified by the radiologist was computed. Figure 4.13

shows the automatically obtained results with the proposed method for three images, along with the position of the nipple as marked by the radiologist. As it can be noticed, the algorithm was able to locate the nipple both inside the breast profile (Fig. 4.13a) and on the profile (Figs. 4.13b and c). In cases where the nipple was nearly invisible and the radiologist provided an approximate position, our method was able to identify the nipple position within 10 mm from the manually marked center. This was mainly due to the efficacy of the geometric constrains placed on the PNRA.

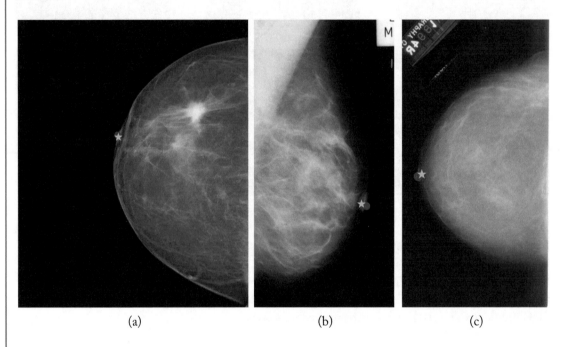

(a)   (b)   (c)

Figure 4.13: Examples of detected nipples with the proposed method, one from each of the three databases of mammograms used in the present study: (a) FFDM image, error= 1.62 mm; (b) mini-MIAS image, error= 1.92 mm; and (c) DDSM image, error = 2.01 mm. Two points are shown on each mammogram corresponding to the nipple position detected automatically (star) and manually marked by the radiologist (circle).

The results of the method were compared with those provided by two other algorithms proposed by Kinoshita et al. [73] and van Engeland et al. [152], using the same dataset of images. Table 4.1 summarizes the comparative analysis. The results of the three methods are slightly better with CC views as compared to MLO views. This is due to the fact that all of the methods depend on accurate segmentation of the pectoral muscle in MLO views. Moreover, variations in the orientation of image acquisition (placement of the compressed breast in relation to the imaging platform) tend to make MLO images prone to alignment errors. Table 4.1 indicates

Table 4.1: Performance of the automated nipple detection method. The accuracy and the outliers are quantified as percentages of images in which the detected nipple location was, respectively, within 10 mm and over 50 mm of the center of the nipple as identified by the radiologist.

Method	View	No. of Images	Mean Error (mm)	Maximal Error (mm)	Detection Accuracy	% of Outliers
The present study	MLO	322	7.29	56.97	77.6%	0.6%
van Engeland et al. [152]	MLO	322	9.51	63.92	69.6%	0.3%
Kinoshita et al. [73]	MLO	322	16.52	88.31	44.4%	3.4%
The present study	CC	244	5.92	49.48	81.6%	0%
van Engeland et al. [152]	CC	244	6.68	32.15	80.7%	0%
Kinoshita et al. [73]	CC	244	8.90	48.73	72.2%	0%
The present study	CC/MLO	566	6.70	56.97	79.3%	0.4%
van Engeland et al. [152]	CC/MLO	566	8.28	63.92	74.4%	0.2%
Kinoshita et al. [73]	CC/MLO	566	13.23	88.31	56.4%	1.9%

the mean and the maximal errors, the detection accuracy, and the number of outliers in the results obtained. The detection accuracy refers to the percentage of images in which the nipple was identified within 10 mm from the manually marked center. Images that had an error larger than 50 mm are defined as outliers, but are included in the derived statistics. The absolute error obtained by our method was, on the average, 6.7 mm over the 566 images processed, with a maximal error of 56.97 mm. Accurate identification of the nipple was achieved for 79.3% of the images. Further comparison of the results of the three methods is provided in Figs. 4.14a and b, as the histogram and cumulative distribution, respectively, of the Euclidean distance between the automatically detected nipple location and the manually marked center. It can be observed that our method outperforms the method by van Engeland et al. [152] and the method by Kinoshita et al. [73]. The number of outliers is slightly higher (only one image) for our method as compared to the results of the method by van Engeland et al. [152], but significantly lower compared to the results of the method by Kinoshita et al. [73]. Out of the 566 mammograms processed, the proposed method failed to detect the nipple in two cases (0.35%), whereas the method by van Engeland et al. [152] and the method by Kinoshita et al. [73] failed to detect the nipple in one case (0.18%) and 11 cases (1.94%), respectively. The two images with the incorrectly detected nipple location are shown in Fig. 4.15. Both failures are due to substantial deformation of the breast in MLO views, which caused a mismatch between the true nipple position and the estimated PNRA.

A direct comparison with the other methods in the literature is not possible due to the use of different databases. However, a comparison in terms of performance is still possible. Yin et al.

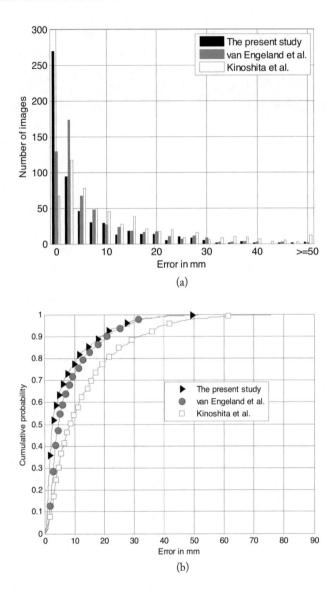

Figure 4.14: (a) Histogram of the Euclidean distance between the detected location of the nipple and the center of the nipple as identified by the radiologist. (b) Cumulative distribution function for the Euclidean distance between the detected location of the nipple and the center of the nipple as identified by the radiologist. Results obtained with the methods by van Engeland et al. [152] and Kinoshita et al. [73] are included for comparison. Reproduced with permission from Casti et al. [18] © Springer.

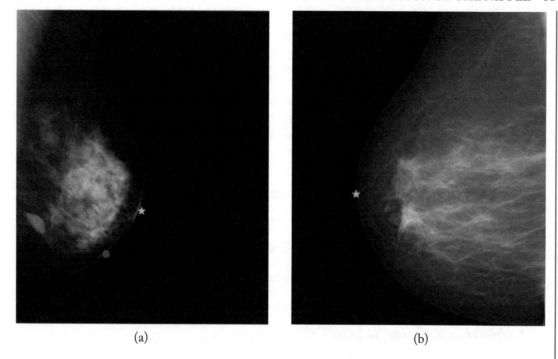

(a)   (b)

Figure 4.15: Two cases of failure of the proposed method: (a) error $=38.72$ mm and (b) error $=33.96$ mm. Two points are shown on each mammogram corresponding to the position of the nipple detected automatically (star) and manually marked by the radiologist (circle).

[164], Méndez et al. [96], and Iglesias and Karssemeijer [63] reported average errors larger than the error of 6.7 mm obtained in our study. The method by Zhou et al. [171], despite having a mean error of 2.5 mm, which is the best result quoted in the literature, results in an unacceptable number of outliers with large errors (2.5% of the total compared to 0.35% obtained in this work). Finally, Chandrasekhar and Attikiouzel [30] tested their method on only 24 mammograms.

One of the strengths of our procedure is that it can detect the nipple even when the nipple is positioned within the breast profile. Another strength is the robustness of the procedure to adverse conditions. The method was tested on a large set of mammograms, including the presence of noise, artifacts, and malignant processes. Experiments showed that the preliminary extraction of the PNRA reduces the chance of false detection of the nipple. Moreover, even when the nipple is nearly invisible on the mammogram, the PNRA provides a reasonable estimate of the nipple position. The main disadvantage of our method is that it requires accurate estimation of the orientation of the pectoral muscle for MLO views to locate properly the PNRA. Substantial distortions of the breast during the imaging procedure can cause failure of the method because of mismatch between the true nipple position and the estimated PNRA. However, improper positioning of the breast during the imaging procedure is also the most frequently encountered problem by radiologists when interpreting mammograms.

4.5 EXTRACTION OF THE BREAST SKIN-LINE

The breast skin-line bounds the area of the mammogram that is of interest for the detection of abnormalities. Several important tasks in computerized analysis of mammograms are strongly dependent on proper and reliable segmentation of the breast region, such as:

- removal of labels and artifacts in SFMs;

- detection of the nipple as a stable reference point on a mammogram;

- extraction of the fibroglandular disk to estimate breast density;

- multiple-view analysis of mammograms for improved detection efficacy;

- multimodality analysis of a lesion's characteristics;

- alignment and subtraction procedures for bilateral comparison of mammograms; and

- registration of mammograms for CBIR.

Nonetheless, many inherent difficulties are encountered in developing automated systems which are capable of working correctly in every situation, some of which are listed below.

- The low contrast of the breast tissue near the skin-air boundary, which is mostly adipose, and appears noisy and dark.

- The tissue superimposition caused by 2D projection that may produce ambiguity in the image.

- The presence of noisy areas and artifacts that compromises image quality.

4.5.1 OVERVIEW OF THE METHODS

The algorithm for the extraction of the breast skin-line [19] combines the detection of a skin-air ribbon obtained by Otsu's thresholding method [107] and the Euclidean distance transform (*DT*) [92] with edge detection by means of multidirectional Gabor filtering. Figure 4.16 illustrates a flowchart of the whole procedure for identification of the breast skin-line. In order to overcome limitations related to the intrinsic variability between mammograms, a novel adaptive values-of-interest (VOI) logarithmic transformation is applied to the original image. The gray-level mapping is specifically designed to be adaptive and to enhance the skin-air interface on mammograms from different modalities, facilitating the detection of the breast boundary in the presence of variations in terms of contrast and noise. The image is then filtered with a

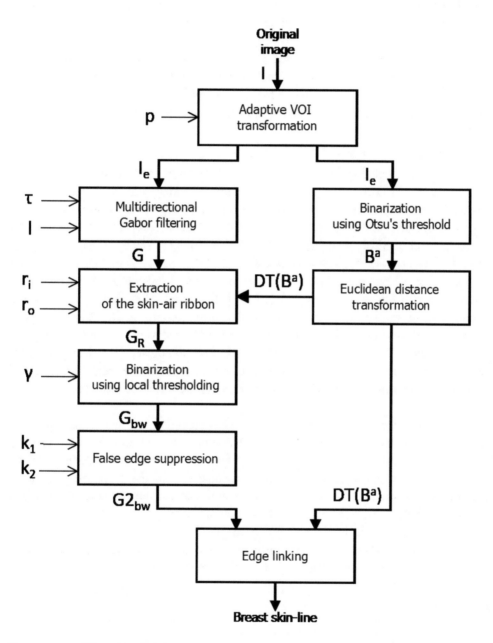

Figure 4.16: Flowchart of the procedure for estimation of the breast skin-line. Reproduced with permission from Casti et al. [19] ©Elsevier.

bank of 18 Gabor filters equally spaced over the range $(-\pi/2, \pi/2]$. Gabor filters are used as edge detectors by means of their magnitude and phase responses. Local constraints are placed to binarize the magnitude response and reduce the number of false edge points, hence confirming only portions of the breast boundary. *DT* is used to propagate the skin-line to the missing portions of the boundary (edge linking). The propagation step is based on the observation reported by Sun et al. [142] that the Euclidean distance from the stroma edge to the actual skin-line is usually uniform. Further details of the method are given in the following sections.

4.5.2   DATASET AND EXPERIMENTAL SETUP

A total of 443 mammographic images, including 249 SFMs from the mini-MIAS database [141] and 194 FFDMs, were used in this study. The images from the mini-MIAS were automatically cropped in order to eliminate the black borders present on both sides of the matrix. Only those images that were not corrupted by the digitization process were retained and used in this work. A set of SFMs, consisting of the same 82 images used by Ferrari et al. [50], Sun et al. [142], Silva et al. [135], and Kus and Karagoz [79], was used in this study to determine the optimal set of parameters of the algorithm and for comparison. Manually drawn breast boundaries provided by an expert radiologist [50] (Radiologist 1 or *R*1) are available for this set of images. Cross-validation of results is provided with reference to *R*1. A second set of mammograms, consisting of 167 SFMs and 194 FFDMs, was used as an independent test set to validate the procedure. A second radiologist, referred to as Radiologist 2 (*R*2), independently marked the breast skin-line on all of the images used in this work with a discrete number of points (15 points per mammogram, on average), which were used as a second reference in the test phase.

Images were normalized to the range $[0, 1]$ and downsampled to a spatial resolution of 300 μm/pixel using a bicubic interpolation method that assigns to each of the output pixels a weighted average of pixel values in the nearest four-by-four neighborhood. The downsampling procedure was applied in order to reduce the computational cost of the proposed procedure. Previous experiments showed that this step does not affect the results substantially while reducing the computational time, especially during the cross-validation process. The automatically extracted contours obtained with the downsampled images were mapped to the original resolution for evaluation and display. Then, after automatically determining the laterality of each image, left-breast mammograms were mirrored about the vertical axis.

4.5.3   METHODS

Adaptive VOI Logarithmic Transformation

In the DICOM standard, various VOI transformations are described for application to projection-based X-ray images. Converting the modality's pixel values into meaningful pixel values, such transformations present the image in a form more suitable for diagnostic interpretation and enhance the perception of the breast tissue components in FFDMs. The VOI transformations can be linear or nonlinear. When the transformation is linear, it refers to a ramp and is determined by two windowing parameters: the window center w_c and the window width w_w. When the transformation is not linear, an intensity mapping can be either explicitly conveyed in a lookup table (LUT), or analytically expressed by means of a bivariate sigmoid function of w_c and w_w as follows:

$$\tilde{I}(x, y) = \frac{1}{1 + \exp\left[-4\dfrac{I(x, y) - w_c}{w_w}\right]},\qquad(4.7)$$

where $I(x, y)$ is the pixel intensity at (x, y) of the input image and $\tilde{I}(x, y)$ is the pixel intensity after VOI transformation. Note that this transformation is similar to the Hurter-Driffield curve that expresses the relation between density and exposure in SFMs. Hence, the use of a VOI sigmoid transformation, instead of a ramp, gives the digital mammogram a film-like aspect. The parameters w_c and w_w are either encoded as data with the DICOM image or defined interactively by the user. In the present algorithm, the parameters w_c and w_w are automatically determined via the gray-level distribution of each mammogram, as described in the following paragraphs.

For our purposes, two main aspects need to be addressed: the dynamic range of the mammogram needs to be compressed and the breast skin-line emphasized; images acquired with different modalities and having different properties need to be processed in the same way. The objectives mentioned above are accomplished as follows. First, the inverse of the sigmoid VOI transformation in Eq. (4.7) is applied to the given image $\tilde{I}$, obtaining the following relation:

$$I(x, y) = \begin{cases} 0 & \text{if } \tilde{I}(x, y) \leq I_0, \\ 1 & \text{if } \tilde{I}(x, y) \geq I_1, \\ w_c - \dfrac{w_w}{4} \log\left[\dfrac{1}{\tilde{I}(x, y)} - 1\right] & \text{otherwise,} \end{cases} \qquad(4.8)$$

where I_0 and I_1 correspond to the values of the VOI sigmoid function in Eq. (4.7) when $I(x, y) = 0$ and $I(x, y) = 1$, respectively, and given by the following expressions:

$$I_0 = \frac{1}{1 + \exp(4w_c/w_w)},\qquad(4.9)$$

$$I_1 = \frac{1}{1 + \exp\left[-4(1 - w_c)/w_w\right]}.\qquad(4.10)$$

The parameters w_c and w_w are adaptively determined after analysis of the gray-level distribution in the given mammogram. A selected cluster of pixels is extracted first as follows. For MLO views, pixels belonging to the pectoral muscle are excluded from the rest of the mammogram, applying the method proposed by Ferrari et al. [49]. For either CC or MLO views, only pixels whose values are higher than the mean intensity of the whole mammogram are considered. The region with the maximum area is selected and represented in terms of its histogram (see Fig. 4.17b). The width w_w^p and the center w_c^p of the confidence interval for a given coverage probability p are computed, and the parameters w_c and w_w are derived according to the following relationships:

$$w_c = w_c^p + 2\sigma_w, \tag{4.11}$$

$$w_w = w_w^p \left(1 + \frac{1}{\sigma_w}\right), \tag{4.12}$$

where σ_w is the standard deviation of the selected distribution. The value of σ_w is used as a corrective term added in order to move the window adaptively toward those intensity values whose dynamic range needs to be compressed, relating the transformation to the variance of the selected distribution. Figure 4.17d illustrates the shape of the inverse sigmoid function, for a coverage probability $p = 0.90$, providing graphical interpretation of the two parameters w_c and w_w. In the example in Fig. 4.17, $w_c = 0.84$ and $w_w = 0.41$, so that $I_0 = 2.93\ e^{-04}$ and $I_1 = 0.82$. As can be noticed, the dynamic range of the high-intensity pixels is compressed either by saturation or by the linear portion of the sigmoid function, while the dynamic range of the most significant pixel values is enlarged. In fact, because the shape of the sigmoid function is determined adaptively, it fixes the characteristic of the obtained image in terms of contrast properties (slope of the linear part) and useful exposure range (length of the linear part), compensating for the variability between mammograms. Figure 4.17c shows the result obtained for the image *mdb007* after the adaptive VOI transformation in Eq. (4.8). Note that the structure of the nipple, which is nearly invisible on the original image (see Fig. 4.17a), is outlined after the VOI transformation along with the breast skin-line.

Multidirectional Gabor Filtering

As described in Section 3.2.2, the multidirectional approach by means of real Gabor filters can be used to detect structures oriented at different angles, and to relate them with the information regarding their orientation. Such information is used in the present algorithm as an attribute to select only the relevant structures in the image, exploiting the magnitude and phase responses of the filters. Preliminary, a high-pass filter defined as unity minus a Gaussian low-pass kernel function, with standard deviation equal to the standard deviation of the Gaussian envelope of the Gabor kernel oriented along the y direction, is applied to the enhanced image in order to accentuate the skin-air interface. Then, 18 real Gabor filters equally spaced over the angular range $(-\pi/2, \pi/2]$ are used as edge detectors.

Figure 4.18c shows the magnitude response obtained for image *mdb*003 of the mini-MIAS database. The original and the enhanced images are shown, respectively, in Figs. 4.18a and b.

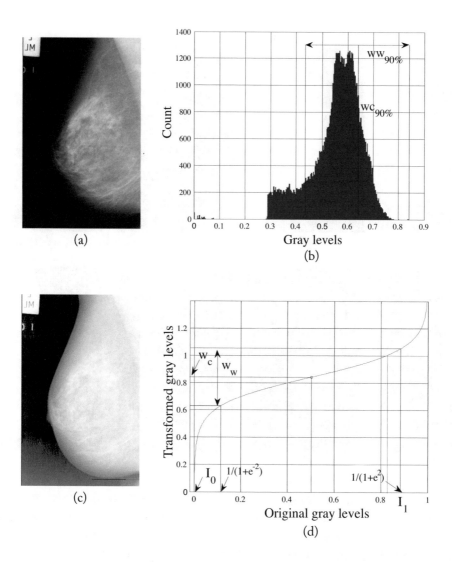

Figure 4.17: (a) Original image *mdb*007 from the mini-MIAS database [141]. (b) Selected gray-level distribution. (c) Image after the adaptive VOI logarithmic transformation in Eq. (4.8). (d) Adaptive VOI function in Eq. (4.8). Reproduced with permission from Casti et al. [19] ©Elsevier.

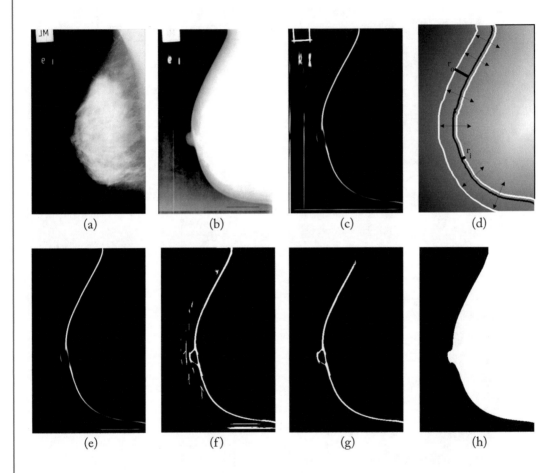

Figure 4.18: (a) Original image (I) mdb003 from the mini-MIAS database [141]. (b) Enhanced image (I_e) obtained with the adaptive VOI logarithmic transformation in Eq. (4.8). (c) Gabor magnitude response (G). (d) Approximate breast boundary B^a (black line) and boundary of the skin-air ribbon G_R (white lines) superimposed on the scaled version of B^a_{DT}. (e) Gabor magnitude restricted to the skin-air ribbon. (f) Gabor magnitude after local thresholding (G_{bw}). (g) Result after suppression of false edge points ($G2_{bw}$). (h) Final binary mask of the breast region. Reproduced with permission from Casti et al. [19] ©Elsevier.

Extraction of the Skin-air Ribbon and Local Thresholding

The initial detection of skin-line points on the Gabor magnitude response $G(x, y)$ is performed in a restricted portion of the image, which is referred to as the skin-air ribbon G_R, localized in the neighborhood of an approximate breast boundary. Given the enhanced image $I(x, y)$ according to Eq. (4.8), the initial boundary is extracted by global histogram thresholding using Otsu's method [107]. Only the region with the maximum area is considered, producing the binary image $B^a(x, y)$ (see the black contour in Fig. 4.18d). Then, DT is applied to $B^a(x, y)$ by assigning to each pixel the Euclidean distance between that pixel and the nearest nonzero pixel of $B^a(x, y)$. An image $B^a_{DT}(x, y)$, whose scaled version is shown in Fig. 4.18d, is obtained as follows:

$$B^a_{DT}(x, y) = DT[B^a(x, y)] - DT[1 - B^a(x, y)] + B^a(x, y) - 0.5, \qquad (4.13)$$

so that positive values of $B^a_{DT}(x, y)$ indicate points within the breast and negative values indicate background points. The term $B^a(x, y) - 0.5$ is added so that, for example, $B_{DT} = \pm 0.5$ for pixels adjacent to the contour. A skin-air ribbon $R(x, y)$ can now be defined using the following relation:

$$R(x, y) = \begin{cases} 1 & \text{if } -r_o < B^a_{DT}(x, y) < r_i, \\ 0 & \text{otherwise,} \end{cases} \qquad (4.14)$$

where r_i and r_o are, respectively, the internal and external extension of the ribbon with respect to the boundary of $B^a(x, y)$, as shown in Fig. 4.18d. Let us denote with S the region such that $R(x, y) = 1$. The pixels of $G(x, y)$ outside S are set to 0, producing $G_R(x, y)$ (see Fig. 4.18e). Then, the resulting image is binarized using a local threshold th determined as follows:

$$th = \gamma \times \max_{(x,y) \in S} [G_R(x, y)], \qquad (4.15)$$

where $\gamma \in [0, 1]$ represents the percentage of the maximum Gabor magnitude response in S, yielding the binary image G_{bw} shown in Fig. 4.18f. As can be seen, G_{bw} contains $N \geq 1$ disjoint areas, $e_i, i = 1, 2, \ldots, N$, of possible edge candidates. In the following section, a procedure for suppression of false edge points is presented.

Suppression of False Edge Points

By knowing the characteristics of the breast boundary, it is possible to define geometrical constraints with the aim of reducing the number of false edge points in the binary image G_{bw}. Given G_{bw}, a false edge is defined by the following points:

- vertical edge points in the right-hand half of the image;

- points belonging to cross-shaped structures; and

- points belonging to relatively small structures.

The conditions stated above were derived by observation of the false edge candidates in images of the Gabor magnitude response after thresholding. The points that meet the stated conditions are detected by exploiting the phase and magnitude responses of Gabor filters, as follows. First, pixels that belong to the horizontal and vertical structures of the mammogram are identified. Because the edge directions are quantized using 18 angles, two ranges of directions are defined over which an edge is considered to be horizontal or vertical. If the phase of a point is in the range of directions $[8\pi/18, \pi/2]$, the point is considered as belonging to a vertical edge (p^v). If the point is in the range of directions $[-\pi/18, \pi/18]$, it is assumed that it belongs to a horizontal edge (p^h). Given the N edge candidates e_i, with $i = 1, 2, \ldots, N$, the related areas a_i, with $i = 1, 2, \ldots, N$, can be computed as the number of pixels of each edge candidate e_i. Then, the procedure for suppression of false edge points is implemented, using two positive constants, $s_1 < 1$ and $s_2 < 1$, as follows.

1. Set $p^v = 0$ in the right-hand half of the image (point suppression).

2. If $\#\,(p^v)_{e_i} + \#\,(p^h)_{e_i} > s_1 \times \max_i a_i$, set $e_i = 0$ (candidate suppression).

3. If $a_i > s_2 \times \max_i a_i$, set $e_i = 0$ (candidate suppression).

Conditions 1 and 2 eliminate pixels that belong to vertical, horizontal, or cross-shaped structures, whose presence is mainly due to noise and artifacts. Finally, condition 3 eliminates those edge points that are relatively small when compared to the edge candidate with the maximum area in the thresholded image. The result of the procedure for suppression of false edge points is a binary image, denoted as $G2_{bw}$, containing $K \leq N$ confirmed edges $e_i, i = 1, 2, \ldots, K$, as shown in Fig.4.18g. In some cases, due to discontinuities in gray-level values caused by substantial alteration in image quality, one or more parts of the contour could be missing. In such images, a linking algorithm, such as that described by Gonzales and Woods [55], is used to assemble skin-line portions into the breast region boundary.

Edge Linking
If the confirmed edge points yield the definition of a closed contour when assembled with the upper, lower, and right-hand-side parts of the image, edge linking is not required. Otherwise, the binary image $G2_{bw}$ is processed as follows.

1. Scan each row of $G2_{bw}$ and detect the first $(p^s_{e_i})$ and the last $(p^f_{e_i})$ points of each of the K confirmed edges e_i.

2. If $K \geq 1$ consider the gap between the edge e_i and the edge e_{i+1} and estimate the gap-boundary distance $D^B_{e_i\,e_{i+1}}$ using B^a_{DT} as $D^B_{e_i\,e_{i+1}} = [B^a_{DT}(p^f_{e_i}) + B^a_{DT}(p^s_{e_{i+1}})]/2$.

3. Generate the intermediate missing portions of the boundary. Scan the row of $G2_{bw}$ from $p^f_{e_i}$ to $p^s_{e_{i+1}}$, and fill (set to 1) the gap in each point p^* that satisfies $B^a_{DT}(p^*) = D^B_{e_i\,e_{i+1}}$; set $G2_{bw}$ to 0 otherwise.

4. Generate the upper and lower missing portions of the boundary. Scan the row of $G2_{bw}$ from 1 to $p^s_{e_1}$, and fill (set to 1) the gap in each point p^* that satisfies $B^a_{DT}(p^*) = B^a_{DT}(p^s_{e_1})$; set $G2_{bw}$ to 0 otherwise. Repeat the process from $p^f_{e_k}$ to the last row of $G2_{bw}$ and set $p^* = 1$ if $B^a_{DT}(p^*) = B^a_{DT}(p^f_{e_k})$.

Once all vertical gaps between confirmed edge pixels are linked, a binary mask is obtained by filling the background from the right-hand edge of the image to the confirmed set of edge pixels. With this additional step, all the horizontal gaps are also filled and a binary image of the breast region is obtained. An additional final step is required to cope with special cases when the curvature of the breast contour in the lower part of the image is such that scanning the image line by line, two edge candidates are detected.

5) Select the point $p^f_{e_K}$ of the last confirmed edge e_K and its coordinates (x_K, y_K). Set to 0 each point p^* with the coordinates (x, y), $x < x_K$ and $y > y_K$, that satisfies $B^a_{DT}(p^f_{e_K}) \leq 0$.

In this way, the curvature of the breast in the lower part of the image is recovered and, finally, the breast skin-line is derived as the boundary of this region. An example is shown in Fig. 4.18h.

Evaluation Methods

When manually drawn boundaries are available for evaluation of results, the automatically detected boundaries can be compared with them by using measures of similarity, such as completeness, correctness, and quality, defined as follows:

$$Completeness = TP/(TP + FN). \tag{4.16}$$

$$Correctness = TP/(TP + FP). \tag{4.17}$$

$$Quality = TP/(FN + TP + FP). \tag{4.18}$$

For each image, TP is the number of pixels that the automatic method assigned to the breast region in agreement with the manually segmented breast region. FN is the number of pixels that the automatic method assigned to the background of the mammogram while the radiologist marked the same as belonging to the breast region. FP is the number of pixels that the automatic method assigned to the breast region while the radiologist assigned the same to the background of the mammogram. These similarity measures reflect previous evaluation criteria used in related works [50, 79, 135, 142]. However, they are difficult to interpret meaningfully because they improve not only with the correctly detected pixels along the contour of the breast, but also with all the correctly detected pixels inside the breast region. In this work, these measures have been included to facilitate comparison with the cited previous works, when the proposed algorithm is applied on the same set of images and evaluated using the same contours provided by $R1$. Given that, for the same test images, the number of positive pixels truly representative of the breast region is the same, such comparison will be meaningful.

When two point sets need to be compared, distance measures are more appropriate and can reflect better the performance of the proposed approach. Given two point sets $X = \{x_1, x_2, \ldots, x_n\}$ and $Y = \{y_1, y_2, \ldots, y_m\}$, belonging to the two boundaries to be compared, different distance measures can be evaluated, as follows.

The Hausdorff distance (D_H) [74] is defined as the maximum among the shortest distances between points x and y:

$$D_H(X, Y) = \max_{x \in X}\{\min_{y \in Y}[d\,(X, Y)]\}, \tag{4.19}$$

where $D_H(X, Y)$ is the Hausdorff distance and $d\,(X, Y)$ is the Euclidean distance.

Another distance measure, referred to as PDM [143], computes the average polyline distance between the two polygons P_X and P_Y connecting the points in X and Y, respectively. The polyline distance of point x_i to the boundary Y is defined as follows:

$$d(x_i, Y) = \min_{s_y \in sides\, P_Y} d(x_i, s_y), \tag{4.20}$$

where $d(x_i, s_y)$ is the shortest distance from the point x_i to the sides s_y of P_Y using Euclidean geometry. For a complete description of PDM, see Suri et al. [143]. Note that the computation of the two distance measures mentioned above from both contours is not appropriate when a significant difference in the cardinality of the two sets of points is present. In fact, because the number of discrete points marked by $R2$ is significantly lower than the number of automatically extracted points, the distances from the automatically extracted contour will be considerably affected by sampling error.

4.5.4 RESULTS AND DISCUSSION

Evaluation of the VOI Transformation

The VOI transformation proposed in this work to optimize the segmentation of the breast skin-line was compared with the logarithmic operation as applied by Ferrari et al. [50]. Results obtained by randomly subsampling 41 of the 82 mammograms from the MIAS database and applying both transformations separately were similar in terms of the average D_H of the automatically detected contours from the manually drawn breast boundaries.

Training, Optimization, and Comparative Analysis

The algorithm for the extraction of the breast skin-line involves the use of several parameters. Before evaluating the performance of the method, these parameters were empirically determined by running experiments with a small subset of training images. The same dataset as used by Ferrari et al. [50], Sun et al. [142], Silva et al. [135], and Kus and Karagoz [79], consisting of 82 mammograms from the mini-MIAS database [141] along with the manually drawn breast boundaries traced by $R1$ was analyzed to determine the optimal parameters of the algorithm. This experiment was performed to facilitate comparison of the results of the present study with those of the studies mentioned above. The average D_H of the automatically detected contours from the manually drawn breast boundaries was used as the criterion to find the most effective set of parameters for the proposed approach.

Experiments were conducted by allowing the eight parameters in our method, summarized in Table 4.2, to range independently over predetermined acceptable intervals of values with fixed incremental steps. The optimal values for the eight parameters, summarized in Table 4.2,

Table 4.2: Values of the optimized parameters used in the proposed algorithm

Parameter	Description	Value
p	Coverage probability of the VOI transformation	0.92
τ	Thickness of the Gabor filter	8.5 *pixels*
l	Elongation of the Gabor filter	2.5 *pixels*
Υ	Threshold factor of the Gabor magnitude response	0.07
r_i	Internal extension of the ribbon	6 *pixels*
r_o	External extension of the ribbon	40 *pixels*
s_1	False edge point suppression coefficient	0.6
s_2	False edge point suppression coefficient	0.3

were determined at the minimum, over the entire set of possible values, of the average D_H over the 82 images. Figure 4.19 shows the average D_H analyzed as a function of the two parameters τ and l (thickness and elongation) of the Gabor filters (Fig. 4.19a) and as a function of the extension, internal (r_i) and external (r_o), of the ribbon (Fig. 4.19b). For the purpose of visualization, the remaining parameters are fixed to the optimal obtained values illustrated in Table 4.2. The final optimal combination of the two parameters is located on the 3.31 mm isodistance contour of each map. The performance measures obtained with the proposed approach are provided in Table 4.3 together with those reported by a few other state-of-the-art methods [50, 79, 135, 142], using the same dataset of images and with respect to the same ground-truth boundaries. The value of D_H obtained in the present study was, on average, 3.31 mm over the 82 mammograms, which is slightly higher than that obtained by Kus and Karagoz [79]. The average PDM was 0.31 mm, while the measures of completeness, correctness, and quality were, respectively, equal to 99.8%, 99.5%, and 99.3%. Our method outperforms the others listed in terms of three of the five performance measures. Although our parameters are optimal, based on the chosen performance criterion, this is consistent with the evaluation performed by the other works that did not evaluate the results obtained with an independent test set of images. A comparison of the results obtained with the proposed method and the deformable model proposed by Ferrari et al. [50] is shown in Fig. 4.20 for image *mdb*003, together with the boundary manually drawn by $R1$. In this example, the proposed method outperforms the deformable model, especially in the upper and lower portions of the breast skin-line and the area of the nipple where the contrast is lower.

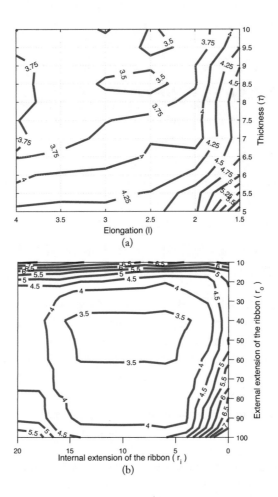

Figure 4.19: Two examples of topographic maps of D_H for various values (a) of elongation (l) and thickness (τ), and (b) of internal (r_i) and external (r_o) extension of the ribbon. The remaining parameters, for the purpose of visualization, are fixed at their optimal values listed in Table 4.2. Reproduced with permission from Casti et al. [19] ©Elsevier.

Table 4.3: Comparative analysis using the same dataset of images and with reference to the same manually drawn boundaries

Method	D_H (mm) $\mu \pm \sigma$	PDM (mm) $\mu \pm \sigma$	Completeness (%)	Correctness (%)	Quality (%)
The present study	3.31 ± 2.39	0.31 ± 0.15	99.8 ± 0.03 E^{-1}	99.5 ± 0.04 E^{-1}	99.3 ± 0.05 E^{-1}
Kus and Karagoz [79]	2.19	0.35	100	98.7	99
Silva et al. [135]	–	0.41 ± 0.16	–	–	–
Sun et al. [142]	–	0.66 ± 0.43	–	–	–
Ferrari et al. [50]	4.49	0.98 ± 0.38	99.4	97.6	97

Testing and Validation Independent of Imaging Modality and Radiologist

In this study, in order to test the consistency of the algorithm's performance and verify its robustness with a large and diversified dataset, the performance measures using independent test sets of images, consisting of mammograms obtained with different modalities (SFM and FFDM) was also evaluated. The test set includes 167 mammograms from the mini-MIAS database (different from those in the training set) and 194 FFDMs. Evaluation of the results was performed with reference to the boundary points marked by $R2$ and identified by the asterisks in Figs. 4.21a, 4.21c, 4.21e, and 4.21g. When computed on the same set of images but with reference to different radiologists (see Fig. 4.22), D_H and PDM result in different average values. With reference to $R2$, the obtained D_H and PDM are, on average, 2.60 mm and 0.88 mm, respectively. As emphasized in Fig. 4.22, PDM measures with reference to $R2$ show a general deterioration of performance, which was expected considering the nature of the ground-truth provided by $R2$. For this reason, the other measures of similarity are not reported in this work. Four examples of the obtained results are shown in Fig. 4.21 along with the ground-truth provided by the two radiologists. Note that only ground-truth points provided by $R2$ were available for mammograms belonging to the test set.

Table 4.4 outlines the results obtained using independent training and test sets along with the confidence intervals with a 0.95 coverage probability. The obtained average values of D_H demonstrate the robustness of the method when an independent set of images is tested using the optimized parameters. Average values of 3.90 mm and 1.95 mm were obtained, respectively, for SFMs and FFDMs. Despite the overestimation, the low average values of PDM (1.08 mm for SFMs and 0.63 mm for FFDM) indicate the efficacy of the method. Figure 4.23 shows the

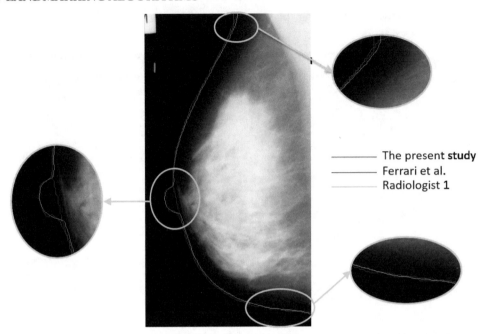

The present **study**
Ferrari et al.
Radiologist **1**

Figure 4.20: Comparison of the results obtained for image $mdb003$ by the skin-line detection method proposed in this work, the method proposed by Ferrari et al. [50] and the boundary manually drawn by $R1$.

Table 4.4: D_H and PDM obtained in relation to the manually marked points provided by $R2$ on different datasets of mammograms using independent training and test sets. N = number of images. $I_{95\%}$ = 95% confidence interval.

Database		N	DH (mm)		PDM (mm)	
			$\mu \pm \sigma$	$I_{95\%}$	$\mu \pm \sigma$	$I_{95\%}$
Train	MIAS	82	2.60 ± 1.68	[0.80 − 6.69]	0.88 ± 0.52	[0.32 − 1.47]
Test	MIAS	167	3.90 ± 4.32	[0.80 − 16.01]	1.08 ±1.16	[0.25 − 2.01]
	FFDM	194	1.95 ± 1.16	[0.78 − 4.00]	0.63 ± 0.15	[0.30 − 0.86]
	Both	361	2.85 ± 3.21	[0.78 − 7.69]	0.84 ± 0.83	[0.25 − 1.04]

cumulative distribution functions obtained for D_H (Fig. 4.23a) and PDM (Fig. 4.23b) between the automatically estimated skin-line and the ground-truth marked by $R1$ and $R2$. In the case of SFMs, better results were obtained for the set of images with which the parameters were optimized than for the test set. In the test phase, the best performance was obtained with FFDMs. This is due to better quality of the images in terms of contrast, noise, and artifacts. In fact, most

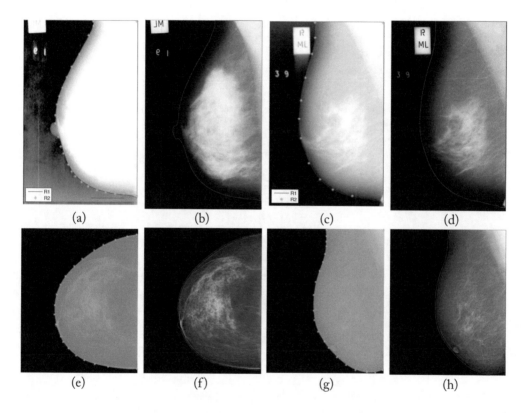

(a)   (b)   (c)   (d)

(e)   (f)   (g)   (h)

Figure 4.21: Results obtained for images (a,b) *mdb*003 and (c,d) *mdb*042 from the mini-MIAS database and for images (e,f) *CD10_PA3_ST1_RIGHT_CC* and (g,h) *CD11_PA2_ST1_LEFT_MLO* from San Paolo Hospital of Bari. Hand-drawn boundary (a,c) from *R*1 and ground-truth points (a,c,e,g) from *R*2 superimposed on the enhanced images. (b,d,f,h) Breast skin-line detected automatically, superimposed on the original image. Reproduced with permission from Casti et al. [19] ©Elsevier.

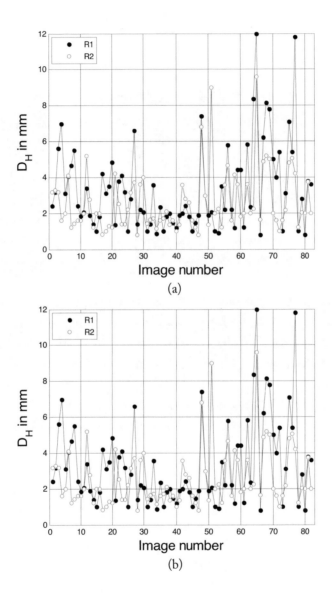

Figure 4.22: (a) D_H and (b) *PDM* obtained with the proposed method relative to the ground-truth provided by radiologists *R*1 and *R*2 for the 82 images from the mini-MIAS database. Reproduced with permission from Casti et al. [19] ©Elsevier.

of the failures of the proposed approach are due to the presence of artifacts and labels in SFMs when they partially overlap the image and could not be completely removed.

Effects of Tissue Composition on the Results

In order to evaluate the effect of tissue composition on the results provided by the proposed approach, Table 4.5 summarizes the obtained distance measures grouped by BI-RADS category. Only images from the MIAS database included the information on the BI-RADS tissue composition and are therefore used for the present analysis. The training set of 82 images and the test set of 167 images from the MIAS database [141] are mutually independent sets. The results show that the variability of D_H and PDM among the different BI-RADS categories of mammograms is not substantial.

Table 4.5: Comparative analysis including BI-RADS categories, using the same dataset of images and with reference to the same manually drawn boundaries. The training set of 82 images and the test set of 167 images from the mini-MIAS database [141] are mutually independent sets.

Breast Composition	Dataset	Number of Images	D_H (mm) $\mu \pm \sigma$	PDM (mm) $\mu \pm \sigma$
1 BI-RADS	R1 trainMIAS	6/82	1.95 ± 0.99	0.25 ± 0.06
	R2 trainMIAS	6/82	2.75 ± 3.07	1.25 ± 1.57
	R2 testMIAS	33/167	4.48 ± 5.03	1.37 ± 1.62
2 BI-RADS	R1 trainMIAS	38/82	3.34 ± 2.40	0.31 ± 0.13
	R2 trainMIAS	38/82	2.29 ± 1.71	0.77 ± 0.32
	R2 testMIAS	69/167	3.83 ± 4.61	1.04 ± 1.29
3 BI-RADS	R1 trainMIAS	31/82	3.57 ± 2.65	0.35 ± 0.17
	R2 trainMIAS	31/82	2.93 ± 1.30	0.92 ± 0.31
	R2 testMIAS	44/167	3.53 ± 3.69	0.91 ± 0.54
4 BI-RADS	R1 trainMIAS	7/82	3.21 ± 1.77	0.27 ± 0.14
	R2 trainMIAS	7/82	2.78 ± 1.58	0.96 ± 0.50
	R2 testMIAS	21/167	3.95 ± 3.69	1.09 ± 0.85

Cross-validation

In order to ensure that the algorithm was tested without mismatch in the radiologist who drew the ground-truth and the format of the ground-truth drawn (whereas $R1$ drew the breast boundary as a contour, $R2$ marked discrete points), cross-validation was performed for the 82 images segmented by $R1$ with the following procedure: 41 images were randomly selected and used to optimize the parameters of the algorithm (training set), the remaining 41 images were used

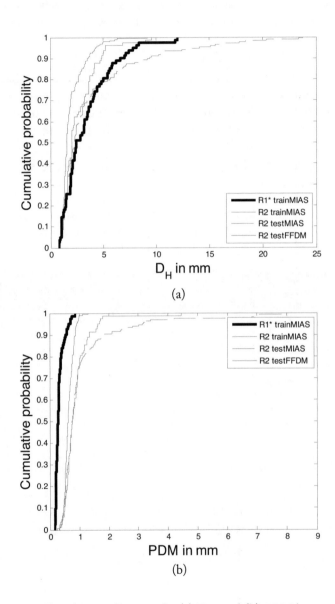

(a)

(b)

Figure 4.23: Cumulative distribution function for (a) D_H and (b) PDM between the automatically estimated skin-line and the ground-truth marked by radiologists $R1$ and $R2$. Note that the parameters were optimized based on the ground-truth provided by $R1$. Reproduced with permission from Casti et al. [19] ©Elsevier.

as the test set to evaluate the performance with the optimized parameters. The procedure was repeated 10 times using independent training and test sets and the results are reported as the average among the training sets and the test sets, in Table 4.6. The results demonstrate that the proposed method achieves good results with reference to $R1$, assuring that the algorithm was tested without mismatch in the radiologist and the format of the ground-truth.

Table 4.6: Results of the cross-validation procedure averaged among 10 repetitions. D_H and PDM were obtained relative to the contours manually drawn by Radiologist 1 using independent training and test sets.

Dataset	N	D_H (mm) $\mu \pm \sigma$	PDM (mm) $\mu \pm \sigma$
Train	41	3.51 ± 0.22	0.37 ± 0.02
Test	41	3.90 ± 0.25	0.42 ± 0.01

4.6   REMARKS

Novel methods for detection of the nipple and for automatic extraction of the breast skin-line were presented in this chapter, together with a modified algorithm for the extraction of the pectoral muscle profile. The main advantages of the proposed procedures are robustness to artifacts and noise, and compatibility with different types of mammographic images, i.e., SFMs and FFDMs from multiple sources, without requiring any type of mapping specific to the imaging protocol (except spatial resolution). When compared with the results obtained by other state-of-the-art methods, the proposed algorithms mostly outperforms the other approaches. In the following chapter, the application of the landmarking algorithms for performing bilateral masking procedures aimed at the detection of bilateral asymmetry is illustrated.

CHAPTER 5

Computer-aided Detection of Bilateral Asymmetry

5.1 PATTERNS OF ASYMMETRY

The existence of bilateral characters in nature provides an invaluable opportunity to understand variability in nature. Bilateral symmetry corresponds to the repetition of parts as mirror images of each other that are located as pairs on the left and right sides of an organism. Symmetry plays an important role with respect to the laws of nature, cosmology, and fundamental physics [57]. Einstein elevated the principles of symmetry to the space-time domain to derive the exact solutions of his field equations of general relativity [35]. Considering the opposite perspective, observed deviations from an ideal state of symmetry can be tracked and correlated with developmental disorders, environmental instabilities, or pathogenic events. With the pioneering studies by Ludwig [85], the subject of asymmetry has been investigated from an evolutionary point of view [109, 110, 151]. Reported works on various patterns of asymmetry observable within populations have led to the definition of three kinds of asymmetry based on analysis of distribution of the population samples: fluctuating asymmetry, antisymmetry, and directional asymmetry [75, 109, 151]. Beyond the specific differences among them, these patterns are believed to measure the ability of the genome to maintain a normal phenotype under stress or imperfect environmental conditions [75, 155]. Palmer and Strobeck [109] postulated the possibility of using deviations from symmetry as an inferential tool for genetic variability. Møller and Swaddle [99] took a step further, suggesting that asymmetry may reveal information on phenotypic and genotypic quality at the individual level. In a recent survey on natural symmetry, Bradshaw and Finlay [13] have stressed the necessity to converge theory and data from diverse fields, bringing symmetry to the inference-making processes. The key point of this discussion is whether patterns of asymmetry as observable manifestations of biological processes can be used to infer on the nature and severity of the processes themselves. Such a question poses important challenges with respect to clinical diagnostic criteria: analysis of the left-right symmetry of the human body may provide meaningful insights to its state of health or disease. Deviations from the ideal state of symmetry of structural components of the body, when present, may suggest the occurrence of diseases at early stages, when more evident signs are not present yet. However, observed deviations from a state of bilateral symmetry which are not definitely of pathological nature but are due, for instance, to age-related, hormonal, or environmental phenomena pose intrinsic difficulties to the analysis of asymmetry and its characterization. Doubtless, additional evidence is required to derive inference from patterns of asymmetry in biological processes, and this is a part of the scope of this work.

5.2 BILATERAL ASYMMETRY IN MAMMOGRAMS

The development of automated methods for quantification of asymmetry as part of a CADe system can facilitate more accurate interpretation of mammograms and assist radiologists in the reporting process, so that the efficacy of breast cancer screening and prevention programs can be improved. The complexity of detecting asymmetry in mammograms lies in finding accurate matches between anatomical structures to be compared and in designing measures capable of distinguishing structural asymmetry from physiological or positioning differences between the two breasts of the same patient.

The identification of all asymmetric findings in a given pair of mammograms is important, because they may be the only clues to breast disease that are detectable on standard mammographic projections, especially when masses, microcalcifications, and/or architectural distortion are not present or visible [134]. Bilateral asymmetry has proved to be an indicator of increased risk of developing breast cancer [59, 133, 170], stressing the importance of special surveillance and follow-up observations of the patients to establish the nature of the asymmetry present.

Radiologists perform comparative studies of the left and right mammograms of the same patient to prevent missing signs of breast disease. When a greater area of tissue with fibroglandular density is detected in a mammogram relative to the corresponding region in the contralateral breast, it is reported as an asymmetric finding, either local or global (see Section 1.1.4 for more details). Asymmetric findings on mammograms may indicate a developing or underlying mass. They can be subtle in presentation and hence overlooked or misinterpreted by radiologists.

The difficulty with the detection of asymmetry arises because the bilateral anomalies caused by a developing or underlying pathological process need to be differentiated from the physiological differences between the two breasts and distortions due to projection artefacts. These confounding factors and subtlety in presentation can cause overlooking or misinterpretation, even by experienced radiologists [88]. Clinical studies have reported that asymmetry accounts for 3–9% of breast cancer cases incorrectly reported by radiologists as showing no evidence of a tumor [15]. Evidence also suggests that asymmetric distribution of fibroglandular density is a common source of FP diagnosis [154].

The development of methods for detection of bilateral asymmetry as part of a CAD system for mammography can serve two purposes. From one side, it can facilitate more accurate interpretation of asymmetric findings by assisting radiologists in the reporting process [54]. On the other side, accurate quantification of asymmetry in mammographic density can be used, in combination with other direct risk factors, such as aging and family history, to optimize screening programs and prevention plans by assessing the near-term individualized risk for having or developing breast cancer [12, 139].

5.3   STATE OF THE ART

Over the last two decades, many researchers have been addressing detection of signs of breast cancer via computerized analysis of mammograms [37, 121, 147]. Some of the reported work used asymmetry between the right and the left mammograms as a strategy to improve the accuracy of CAD systems in detecting masses or microcalcifications [45, 67, 68, 158, 164]. Only a limited number of studies have addressed the detection of asymmetry as a distinctive sign for diagnosis [48, 81, 98, 124, 150, 156, 157].

The earliest work on detection of bilateral asymmetry was performed by Lau and Bischof [81], who defined an asymmetry measure combining descriptors of roughness, brightness, and directionality. The scope of their work was to localize 13 suspicious asymmetric areas on a set of 10 asymmetric pairs of mammograms. Miller and Astley [98] proposed a semiautomated method to classify pairs of mammograms as normal or abnormal using shape, texture, and density features. The procedure consisted of segmentation of mammograms into fat and non fat regions, asymmetry measurements extracted from non fat regions, and classification. In order to obtain a proper match between corresponding areas of the two breasts of a subject, radiologists' annotations were used to segment the non-fat regions of the mammograms. With a set of 47 normal and 28 asymmetric pairs of mammograms, an accuracy of 76% was obtained using linear discriminant analysis (LDA) and levae-one-out (LOO) cross-validation.

The study by Ferrari et al. [48] introduced the application of Gabor wavelets and rose diagrams to quantify differences in oriented textural patterns in mammograms. Using quadratic discriminant analysis (QDA) and LOO cross-validation an accuracy of 74.4% was achieved. In a subsequent work [124], the same authors improved the accuracy of their method to 84.4% by including morphological and density features. Pairs of MLO views from the mini-MIAS database [141] were used for validation of their methods, of which 22 were normal, 14 asymmetric, and 8 contained architectural distortion.

In the work by Tzikopoulos et al. [150], the use of 114 differential features was tested on the whole mini-MIAS database, which includes 15 asymmetric pairs of mammograms; the remaining 146 pairs were treated as being symmetric, despite the presence of masses, microcalcifications, and architectural distortion. The reported accuracy was 84.5% using support vector machines (SVM).

The studies by Wang and colleagues [156, 157] demonstrated the feasibility of applying a computerized scheme, based on the use of asymmetric density features, for the assessment of early signs of breast cancer. Their method was tested on 100 pairs of normal mammograms and 100 pairs of prior mammograms, and achieved an area under the ROC curve (A_z) of 0.78 ± 0.02 using an artificial neural network (ANN) classifier.

Table 5.1 summarizes methods for the analysis of bilateral asymmetry and the corresponding performance reported by previously published work. The state of the art suggests that automatic detection of asymmetry in mammograms can be achieved, but more effort is needed to devise new methods to improve performance levels and to progress toward clinical application.

Table 5.1: Summary of methods and performance statistics for the analysis of bilateral mammographic asymmetry

Authors	Dataset	Methods and Results
Miller and Astley [98]	75 MLO and mediolateral (ML) mammograms including 47 normal and 28 asymmetric pairs.	Six shape, brightness, and topology features extracted from manually marked regions of fibroglandular components and selected using the F-test; LDA classifier and LOO cross-validation: accuracy = 76%.
Ferrari et al. [48]	80 MLO mammograms from mini-MIAS [141] including 20 normal, 14 asymmetric, and six architectural distortion cases.	Gabor wavelets, Karhunen-Loève transform, Otsu's method, rose diagrams, and three statistical features selected using exhaustive combination; QDA classifier and LOO cross-validation: accuracy = 74.4%.
Rangayyan et al. [124]	88 MLO mammograms including 22 normal cases, 14 asymmetric cases, and eight architectural distortion cases from mini-MIAS [141].	Alignment of the phase responses of Gabor wavelets with reference to the corresponding pectoral muscle edges, four directional features selected using exhaustive combination; quadratic Bayesian classifier and LOO cross-validation: accuracy = 84.4%.
Tzikopoulos et al. [150]	All of the 322 MLO mammograms from mini-MIAS [141] including 15 asymmetric cases; the remaining cases treated as symmetric.	Minimum cross-entropy thresholding, 18 differential first-order statistical features selected using the t-test; SVM classifier and LOO cross-validation: accuracy = 85.7%.
Wang et al. [157]	800 randomly selected FFDMs, including CC and MLO views of 100 normal cases and 100 verified positive cases for having high risk of developing breast cancer.	20 selected features including statistical, textural, and density features from automatically extracted ROIs of CC views and the entire segmented breast areas of MLO views; genetic algorithms, ANN classifier, and leave-one-patient-out cross-validation: area under the ROC curve of 0.78.
Tan et al. [146]	Full-eld digital prior mammograms of 645 cases, including CC and MLO views of 362 normal cases and 283 verified positive cases for having high risk of developing breast cancer.	Nine differential density features in addition to woman's age selected using the sequential forward floating algorithm; SVM classifier and 10-fold cross-validation: area under the ROC curve of 0.716.
Only works with accuracy results and/or ROC analysis are listed.		

5.4 OVERVIEW OF THE METHODS

In the following sections, novel methods for computerized detection of bilateral asymmetry are presented. Three different procedures are investigated which share the following main steps: computerized masking of the left and right mammograms of a patient via the algorithms for landmarking described in Chapter 4, quantification of the differences between paired mammographic regions by means of various measures of similarity or dissimilarity, and classification of pairs of mammograms as normal or asymmetric by means of the extracted features. Method 1 presents a novel application of Moran's index to measure the angular covariance between rose diagrams related to the phase and magnitude responses of multidirectional Gabor filters [22]. Method 2 is based on the analysis of spatial correlation of gray-scale values of pixels with reference to the position of the nipple [25]. Method 3 combines spatial and frequency information by using spherical semivariogram descriptors and correlation-based structural similarity indices [24, 27].

5.5 DATASET AND EXPERIMENTAL SETUP

The procedures for detection of bilateral asymmetry were tested by selecting all of the available cases in public databases with reported asymmetry. To allow direct comparisons of the performance obtained in this study with future work and for the inclusion of control cases, an equal number of normal cases in descending order starting from the first available normal case, i.e., case 0002 of volume normal_01 for DDSM and images mdb003 and mdb004 for mini-MIAS, were selected. The dataset of images thus obtained is composed of a total of 188 mammograms (94 pairs), of which 128 (64 pairs) are from DDSM including CC and MLO projections of the two breasts of each subject, and 60 MLO mammograms (30 pairs) are from mini-MIAS. The asymmetric pairs, 47 in total, include 16 pairs of focal asymmetry and 16 pairs of global asymmetry from DDSM, and 15 cases of asymmetry from mini-MIAS database. The images have spatial resolution of 42.5, 43, 50, or 200 and pixel depth of 8, 12, or 16 bits/pixel (bpp). Proven ground truth of the asymmetric findings was available for all the cases. The first two methods were tested only on the DDSM dataset (128 images), while the method based on the analysis of structural similarity was tested and validated on the whole dataset of mammograms (188 images). All images were first downsampled using bicubic interpolation to a spatial resolution of 300 to reduce the computational time of the procedures and also unify the resolution of images from different databases. Due to the mixed nature of the cases used, the present work is aimed at identification of a pair of mammograms or a patient's case with two sets of CC and MLO mammograms as being normal or asymmetric; no specific abnormal region is detected or segmented in a given mammogram.

5.6   TABÁR MASKING PROCEDURES

As described in Section 1.1.6, Tabár introduced guidelines to the use of bilateral masking procedures by radiologists [145, 167]. Following radiologists' criteria for interpretation of mammograms, an automated implementation of Tabár masking procedures is presented. The hypothesis underlying this study is that automated bilateral masking of mammograms can serve computerized systems for detection of structural asymmetry by providing localized and matched regions for quantitative comparison of mammograms [16, 21, 24, 25, 27]. It is assumed that the statistical features of mammograms are spatially nonstationary and that the differences in the structural information to be investigated are also space-variant, thus requiring accurate matching of the areas under investigation.

The procedure, that is oriented to the analysis of breast asymmetries, requires effective matching between the regions of the breast that will be compared at each step. Hence, paired matching points on the mammograms are required. The landmarks, including the breast skinline, the pectoral muscle (in MLO views), and the nipple, detected by using the algorithms described in Chapter 4, were used as anatomical reference structures for masking. Examples of a malignant asymmetric case from DDSM and a normal case from mini-MIAS are given in Figs. 5.1a and b, respectively. The focal asymmetric region is outlined in green for the DDSM case. The detected landmarks are illustrated in Figs. 5.3c–f. The binary masks of the obtained breast regions and linear approximations of pectoral muscle edges are superimposed on the images. The squares correspond to the positions of the nipple.

For each pair of images, segmentation of paired strips was performed to derive corresponding mammographic regions, as follows.

Masking #1: Horizontal (for CC views) and annular (for MLO views) strips were segmented by dividing the line from the top-most to the lowest pixel inside the breast region (for CC views) and the perpendicular line from the nipple to the pectoral muscle (for MLO views) into eight equally spaced segments and annuli, as shown in Figs. 5.1c and d, respectively.

Masking #2: Vertical (for CC views) and oblique (for MLO views) strips were segmented parallel to the chest wall and the pectoral muscle, respectively, by dividing the perpendicular line from the nipple to the chest wall (for CC views) and the perpendicular line from the nipple to the pectoral muscle line (for MLO views) into eight equally spaced segments, as shown in Figs. 5.1e and f.

In order to derive similarity indices (see Section 5.10.2) and facilitate meaningful comparison, rectangular regions were derived by extracting the largest rectangles enclosed in each segmented strip as illustrated in Fig. 5.2, including the whole breast region from which a rectangular central region was derived. The annular strips obtained via masking #1 for MLO views in Fig. 5.1d were substituted by horizontal strips, as shown in Fig. 5.2b. Each segmented strip extracted from a left mammogram was flipped left to right and paired with the corresponding strip in the contralateral mammogram, and used, in addition to the whole breast regions, for feature extraction.

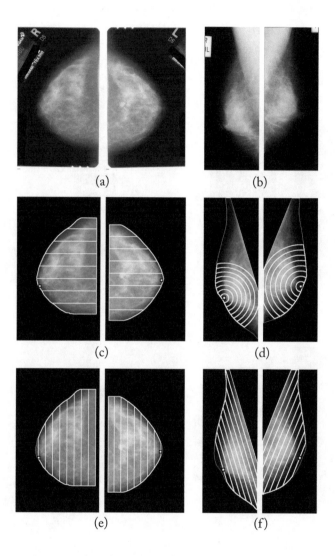

Figure 5.1: Examples of pairs of (a,c,e) malignant asymmetric mammograms, CC views—case *A*-1725-1—from DDSM [61], and (b,d,f) normal mammograms, MLO views—images mdb035 and mdb036—from mini-MIAS [141]. (a,b) Original images. The focal asymmetric regions are outlined in green. (c,d,e,f) Landmarking and bilateral strips obtained via (c,d) masking #1: horizontal/annular masking and (e,f) masking #2: vertical/oblique masking. The binary masks of the obtained breast regions and linear approximations of pectoral muscle edges are superimposed on the images. The squares correspond to the positions of the nipple detected automatically. Reproduced with permission from Casti et al. [27] © IEEE.

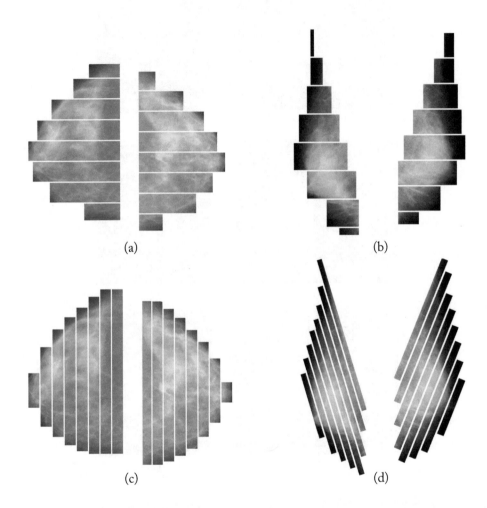

Figure 5.2: Rectangular regions derived by extracting the largest rectangles enclosed in each strip of the masking procedures illustrated in Fig. 5.1. (a,c) Rectangular regions from masking #1: horizontal/annular masking. (b,d) Rectangular regions from masking #2: vertical/oblique masking. Reproduced with permission from Casti et al. [27] © IEEE.

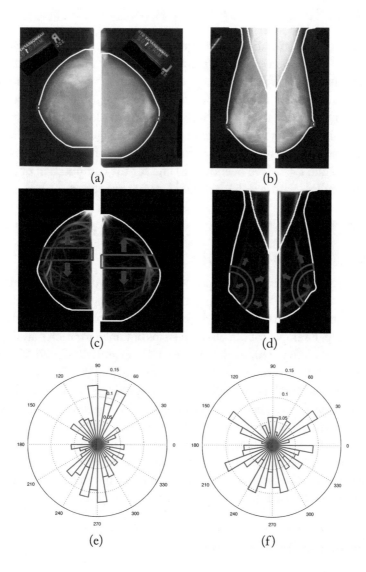

Figure 5.3: Results obtained for a pair of (a,c,e) asymmetric CC and (b,d,f) normal MLO mammograms. (a,b) Original mammograms. The automatically detected breast skin-line and the pectoral muscle line are superimposed on the images. The squares correspond to the positions of the nipple detected automatically. (c,d) Gabor magnitude responses and strips selected at one step of each masking procedure #1: (a) horizontal and (b) annular masking. (e,f) Rose diagrams of the phase response related to the strips selected at one step of each masking procedure. Reproduced with permission from Casti et al. [27] © IEEE.

5.7   EXTRACTION OF DIRECTIONAL COMPONENTS

In addition to the spatial distribution of the gray-scale values, the directional components of breast tissue patterns were investigated to quantify differences in the orientation of the structures of the breast parenchyma. With this purpose, the analysis of phase and structural similarity described in the following sections was performed on the magnitude and phase responses of a set of $N = 18$ equally spaced real Gabor filters. The resulting images of magnitude and phase responses were used, in addition to the original gray-scale images, to derive the paired bilateral strips via masking procedures described in Section 5.6 and used for the analysis of phase similarity with $\tau = 24$ pixels and $l = 3$ (see Section 5.8) and for the computation of correlation-based structural similarity indices with $\tau = 6$ pixels and $l = 8$ (see Section 5.10.2).

5.8   METHOD 1: ANALYSIS OF PHASE SIMILARITY

5.8.1   CALCULATION OF ROSE DIAGRAMS

The images of the magnitude and phase responses obtained for the right and left mammograms of each case were coupled and masked using the four masking procedures described in Section 5.6, so that the information content in the extracted bilateral regions could be analyzed and compared strip by strip. Examples of the bilateral strips obtained for each masking procedure are shown in Figs. 5.3 and 5.4 for a pair of asymmetric CC (a,c,e) and normal MLO (b,d,f) mammograms. The original mammograms are shown in insets (a) and (b). The automatically detected breast skin-line and the pectoral muscle line are superimposed on the images. The squares correspond to the positions of the nipple detected automatically. The contours of the strips extracted via the masking #1 and #2 are superimposed on the Gabor magnitude responses in insets (c) and (d) of Figs. 5.3 and 5.4, respectively. The directional information of each strip was mapped into rose diagrams with 18 sectors, so that each sector is oriented at the angle corresponding to one of the 18 directions (or phases) analyzed. According to the information mapped, two rose diagrams were computed for each bilateral pair of masked strips of mammograms:

1. rose diagram of the phase response, whose sectors are proportional to the number of pixels of the corresponding Gabor phase; and

2. rose diagram of the magnitude response, whose sectors are proportional to the values of the Gabor magnitude of all the pixels corresponding to the related Gabor phase.

The area of each rose diagram was normalized to unity so that the right and left mammograms could be compared. Examples of the rose diagrams of the phase responses corresponding to the selected strips are shown in Figs. 5.3 and 5.4 in insets (e) and (f) with paired colors: red for the right mammogram and blue for the left mammogram.

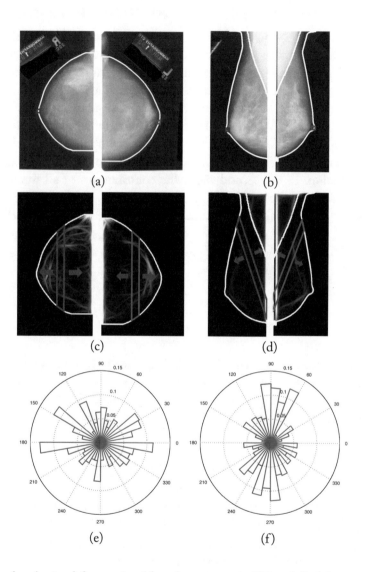

Figure 5.4: Results obtained for a pair of (a,c,e) asymmetric CC and (b,d,f) normal MLO mammograms. (a,b) Original mammograms. The automatically detected breast skin-line and the pectoral muscle line are superimposed on the images. The squares correspond to the positions of the nipple detected automatically. (c,d) Gabor magnitude responses and strips selected at one step of each masking procedure #2: (c) vertical and (d) oblique masking. (e,f) Rose diagrams of the phase response related to the strips selected at one step of each masking procedure. Reproduced with permission from Casti et al. [22] © IEEE.

5.8.2 COMPUTATION OF THE ANGULAR SIMILARITY INDEX

Measures of spatial autocorrelation are used in geostatistics to analyze whether close neighbors over a given surface share similar properties [33]. In the present method, a novel application of Moran's index [100] is investigated by defining a new angular similarity measure to quantify the angular covariance between paired rose diagrams. Given a pair of mammograms and M corresponding bilateral masked strips, computation of the angular similarity index consists of two stages. In the first stage, a Moran-like measure is used to quantify the degree of similarity between the rose diagrams of the right and left directional components in each pair of bilateral strips. The measure has been designed to quantify the degree to which close angles (or sectors) share similar information. In the second stage, the contributions of all the measures derived from the M pairs of strips are pooled into a single score.

Let r_i^m and l_j^m be, respectively, the value of the right i^{th} and the left j^{th} sectors of the rose diagrams related to the mth pair of strips, with $i, j = 1, 2, \ldots, N$, N being the number of sectors of each rose diagram ($N = 18$ in the present work). We compute a Moran-like measure I_m related to the mth pair of strips as follows:

$$I_m = \sum_{i=1}^{N} \sum_{j=1}^{N} w_{ij}(r_i^m - 1/N)(l_j^m - 1/N), \tag{5.1}$$

where w_{ij} is the adjacency matrix that reflects the angular distance between the ith right and the jth left sectors of the rose diagrams. We set $w_{ij} = (|i - j| + 1)^{-1}$, so that w_{ij} reaches its maximum when sectors of corresponding angles are considered. Then, given the M measures related to the M pairs of strips ($M = 8$ in the present work), the proposed angular similarity index is computed as

$$S_A = \sum_{m=1}^{M} I_m. \tag{5.2}$$

Large positive values of S_A indicate higher similarity between the directional components of the right and left breasts, which should correspond to a normal pair of mammograms. Lower values, on the contrary, indicate differences in the bilateral directional structures of the two mammograms and, consequently, the presence of asymmetry.

5.8.3 PATTERN CLASSIFICATION AND CROSS-VALIDATION

The angular similarity index described above was computed from the magnitude and phase rose diagrams resulting from the four masking procedures, so that eight features were extracted, in total, from the mammographic examination results of each patient, two for each of the four masking procedures (#1 and #2 for CC views, and #1 and #2 for MLO views). The ROCKIT package [140] was used to estimate the A_z values (and standard error, SE) for the eight proposed features. The classification of the cases as normal or asymmetric pairs was achieved by means of two classifiers: Fisher linear discriminant analysis (FLDA) and the Bayesian classifier (QDA with the Bayesian assumption, BQDA) [42]. Features were automatically selected in the training sets of each experiment based on the values of A_z. Leave-one-patient-out cross-validation was performed on the whole dataset.

5.8.4 RESULTS AND DISCUSSION

Performance analysis of the angular similarity index is presented in Table 5.2. The results are related to the rose diagrams of the phase and magnitude responses obtained via multidirectional Gabor filtering. In order to evaluate the efficacy of the proposed approach, values of A_z are reported for the four proposed masking procedures and also when the whole breast regions were used without any masking procedure. The highest A_z of 0.8411 (SE = 0.0729) was obtained with medial masking of the Gabor phase response, which is independent of the procedures and results for the extraction of the nipple and the pectoral muscle. This indicates that the accuracy in the detection of the matching structures influences the performance of the angular similarity index. This is also confirmed by the poorer results obtained using MLO views, for which the results of the masking procedures are influenced by the extraction of the pectoral muscle. In particular, A_z values lower than 0.5 imply that higher values of the angular similarity index are correlated with asymmetric pairs, which contradicts the expected trend and, therefore, were not considered reliable. As shown in Table 5.3, the use of FLDA, along with automatic feature selection and the leave-one-patient-out cross-validation procedure, provided higher classification accuracy than BQDA, with A_z (SE) of 0.8435 (0.0729), 0.71929 (0.0925), and 0.8381 (0.0725), obtained, respectively, using CC, MLO, and both CC and MLO views. Sensitivity and specificity rates of 75% and 94% were obtained, respectively, at the operating point identified by the minimum distance between the point (0, 1) and the ROC curve.

   Results of pattern classification demonstrate that the combination of automated masking procedures with the method for the analysis of phase similarity is an effective strategy to detect bilateral asymmetry in mammograms. The results are promising, considering that only one similarity measure was used in this study to classify a given pair of mammograms as being asymmetric or normal. The obtained performance levels have been improved by means of analysis of spatial correlation with respect to the nipple position, which is described in the following section.

5.9 METHOD 2: ANALYSIS OF SPATIAL CORRELATION

5.9.1 COMPUTATION OF MEASURES OF SPATIAL CORRELATION

Each image was further downsampled to 600 pixel and the differences in the spatial distribution of gray-scale levels were quantified as follows. Accurate matching of corresponding mammographic strips was performed by means of the four masking procedures described in Section 5.6. Then, for each strip, measures of spatial correlation, Sc, were computed with reference to the po-

Table 5.2: Performance analysis of the angular similarity index using the rose diagrams related to the Gabor phase and magnitude responses of the right and left mammograms. Symbol * indicates $A_z < 0.5$ and behavior of the angular similarity index contrary to expectations.

View	Masking	A_z (SE)			
		Phase Response		Magnitude Response	
CC	Medial	0.8411	(0.0736)	0.7266	(0.0913)
	Retroglandular	0.8379	(0.0713)	0.7667	(0.0836)
	None	0.5543	(0.1034)	0.7249	(0.0935)
MLO	Milky	0.3994*	(0.1021)	0.6955	(0.0945)
	Retroareolar	0.5398	(0.1047)	0.7344	(0.0908)
	None	0.3202*	(0.0960)	0.6924	(0.0924)

The A_z (and standard error, SE) values were estimated using ROCKIT [140].

Table 5.3: Results of pattern classification in terms of A_z (SE) using leave-one-patient-out cross-validation and two classifiers

	CC	MLO	CC and MLO
FLDA	0.8435 (0.0729)	0.7193 (0.0925)	0.8381 (0.0725)
BQDA	0.7849 (0.0832)	0.6981 (0.0953)	0.8168 (0.0826)

sition of the nipple by comparing a matrix of differences in gray-scale levels, Δf_{ij}, with a matrix of distances, Δd_{ij}, as follows:

$$Sc = \frac{2}{n(n-1)} \sum_{i=1}^{n} \sum_{j=1}^{n} \Delta f_{ij} \, \Delta d_{ij}, \tag{5.3}$$

where n is the number of pixels within each strip. Radial correlation was estimated by assigning $\Delta d_{ij} = |d_i - d_j|$, with d_i, $i = 1, 2, \ldots, n$, equal to the length of the ith pixel's position vector from the nipple. Angular correlation was quantified by assigning $\Delta d_{ij} = \sin \theta_{ij}$, where θ_{ij} is the angle between the position vectors of the ith and the jth pixels. Given the gray-scale level, f_i, of the ith pixel, based on different formulations of Δf_{ij} in Eq. (5.3), four differential spatial correlation features for both angular and radial correlation were defined as follows:

- $\Delta Sc_1 = |Sc_{dx} - Sc_{sn}|$ with $\Delta f_{ij} = |f_i - f_j|$,
- $\Delta Sc_2 = |Sc_{dx} - Sc_{sn}|$ with $\Delta f_{ij} = |f_i - f_j|/(\max_k f_k + \min_k f_k)$,
- $\Delta Sc_3 = |Sc_{dx} - Sc_{sn}|$ with $\Delta f_{ij} = (|f_i - f_j| - \bar{f})/\sigma_f$, and
- $\Delta Sc_4 = |Sc_{dx} - Sc_{sn}|/(Sc_{dx} + Sc_{sn})$ with $\Delta f_{ij} = |f_i - f_j|$,

where dx and sn indicate the right and left mammograms, and $\bar{f}$ and σ_f are the mean and standard deviation of gray-scale values within each strip, respectively. The differential features obtained for each of the eight pairs of strips of a given masking procedure were summed, resulting in 32 features for every patient.

5.9.2 PATTERN CLASSIFICATION AND CROSS-VALIDATION

A threshold of 0.6 on the A_z values of the individual features, followed by stepwise logistic regression was applied to the training set of each leave-one-patient-out experiment for automatic selection of features. Three classifiers—LDA, QDA, and an ANN with radial basis functions (ANN-RBF)—were used for classification of mammograms as normal or asymmetric pairs.

5.9.3 RESULTS AND DISCUSSION

The results of performance analysis of the individual features are presented in Table 5.4. The A_z values are reported for the two mammographic projections and the four related masking procedures. The A_z values obtained when the whole breast regions were used for computation of features without using any masking procedure are also reported. The highest A_z of 0.75 was obtained with the ΔSc_3 correlation measure and medial masking of CC views. The results obtained with MLO views are poorer as compared to the results achieved with CC views. Overall, the masking procedures improve the discriminating ability of the features. The combination of features from both CC and MLO views using LDA, QDA, and ANN-RBF classifiers along with automatic feature selection and the leave-one-patient-out cross-validation provided, respectively, A_z (SE) of 0.83 (0.07), 0.72 (0.09), and 0.87 (0.08). The best classification accuracy obtained was 91%, with sensitivity of 1.0 and specificity of 0.81, using the ANN-RBF classifier.

Table 5.4: Classification performance of individual features for spatial pattern analysis of mammograms. Results are given in terms of A_z. Cases with $A_z > 0.7$ are shown in bold.

View	Masking	Radial Correlation				Angular Correlation			
		ΔSc_1	ΔSc_2	ΔSc_3	ΔSc_4	ΔSc_1	ΔSc_2	ΔSc_3	ΔSc_4
CC	Medial	0.67	0.69	**0.75**	0.55	0.60	0.61	0.57	0.68
	Retroglandular	0.59	0.47	0.52	0.68	0.47	0.39	0.47	0.54
	None	0.41	0.62	**0.72**	0.43	0.46	0.41	0.50	0.57
MLO	Milky	0.36	0.49	0.50	0.45	0.54	0.63	0.57	0.57
	Retroareolar	0.61	0.61	0.65	0.58	0.55	0.59	0.48	0.57
	None	0.55	0.55	0.50	0.54	0.56	0.47	0.52	0.58

The A_z (and standard error, SE) values were estimated using ROCKIT [140].

The obtained results are better than the other results reported in the literature [48, 98, 124, 150, 157] and improve the results obtained with the analysis of phase similarity reported in Section 5.8.4. This analysis of spatial correlation indicates that the differences in the spatial distribution of pixel values within paired mammographic strips computed with reference to the position of the nipple can be used for detection of bilateral asymmetry.

5.10 METHOD 3: ANALYSIS OF STRUCTURAL SIMILARITY

5.10.1 SPHERICAL SEMIVARIOGRAM DESCRIPTORS

Structural variations among pixel values in a given ROI can be quantitatively described by the variogram [65], which is a descriptor that quantifies the degree of spatial dependence between samples. Given $N(h)$ pairs of pixels separated by a distance h, the semivariogram, $\gamma(h)$, is defined as

$$\gamma(h) = \frac{1}{2N(h)} \sum_{m=1}^{N(h)} \left[f(\mathbf{u}_{m,0}) - f(\mathbf{u}_{m,h}) \right]^2, \tag{5.4}$$

where $\mathbf{u}_{m,0}$ and $\mathbf{u}_{m,h}$, $m = 1, 2, \ldots, N(h)$, are vectors of spatial coordinates (x, y) separated by the lag distance h, and f is the gray-scale level at the given spatial locations (see Fig. 5.5a). The strength of this approach lies in characterizing the relationship between the spatial and statistical characteristics of pixels in the image that, in our study, may be descriptive of an underlying pathological phenomenon. Ericeira et al. [45] used the empirical values of variogram and cross-variogram functions computed from pairs of windows of size 32×32 pixels to detect masses in mammograms. In this work, we use semivariogram analysis to derive four spherical semivariogram descriptors.

The images were further downsampled by a factor of five to reduce the computational cost, to an effective resolution of 1.5 mm/pixel. The maximum value of h to be investigated, $h_{\max}$, was set equal to one-half of the maximum distance between pairs of pixels in each region. The range of distances from 0 to $h_{\max}$ was divided into 20 equally spaced bins and pixel pairs were aggregated accordingly to estimate the empirical semivariogram γ_n for representative distances of each aggregate. Least-squares fitting of each empirical semivariogram was performed using a spherical structure function, as

$$\hat{\gamma}(h) = \begin{cases} a + s \left(\frac{3h}{2r} - \frac{h^3}{2r^3} \right), & \text{if } h \leq r, \\ a + s, & \text{if } h > r, \end{cases} \tag{5.5}$$

where a, the *nugget*, represents the discontinuity at the origin due to small-scale variations; s, the *sill*, gives an estimate of the variance of pixels; and r, the *range* of influence of the spatial structure, corresponds to the distance at which the semivariogram stops increasing [106], as shown in Fig. 5.5b.

The spherical semivariogram functions obtained using the whole breast regions are illustrated for the asymmetric malignant case A-1725-1 of DDSM [61] and the normal case

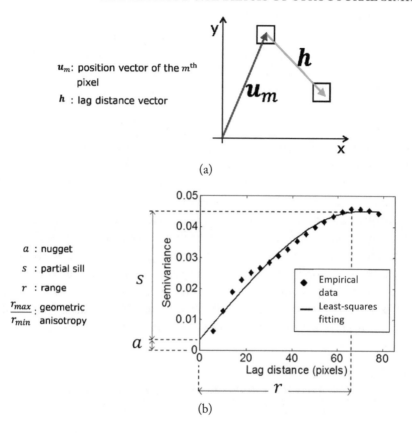

(a)

(b)

Figure 5.5: (a) Vectors of spatial coordinates (x, y) for computation of the semivariogram. (b) Example of empirical semivariogram and the corresponding spherical least-squares fit. The nugget, a, the sill, s, and the range, r, are indicated.

mdb035/mdb036 of mini-MIAS [141] in Figs. 5.6a and b, respectively; the nugget, the sill, and the range are also indicated. Separate anisotropic semivariograms, $\gamma(h_\alpha)$, were also computed for the $N(h_\alpha)$ pairs of gray-scale values separated by the lag distance vector, h_α, oriented at the angle α, with $\alpha = 0°, 30°, \ldots, 180°$, to quantify the behavior of the autocorrelation structures in different directions of analysis. When the structural variations among pixel values are dependent on the direction of analysis, the behavior is referred to as geometric anisotropy, g, and can be quantified by the anisotropy ratio, as the range of the anisotropic semivariogram in the direction producing the longest range divided by the range in the direction with the smallest range. The absolute differences between the four parameters derived from each region of the right mammogram and the corresponding parameters derived from the matching region of the left mammogram, $\Delta V_a = |a_R - a_L|$, $\Delta V_s = |s_R - s_L|$, $\Delta V_R = |r_R - r_L|$, and $\Delta V_g = |g_R - g_L|$, were used as spherical semivariogram descriptors. The differential isotropic semivariogram descriptors obtained for the asymmetric and normal case examples are given in the caption of Fig. 5.6. Higher values of the differential descriptors indicate lower similarity between the paired regions analyzed and are expected to be related to asymmetric cases. The differential values obtained from the eight strips of each masking procedure were summed together [24].

5.10.2 CORRELATION-BASED STRUCTURAL SIMILARITY

The key role of structural information in human image perception has been pointed out by the recent development of the Structural SIMilarity (SSIM) and Complex Wavelet Structural SIMilarity (CW-SSIM) indices for image quality assessment [131, 159]. The structural approach consists of analyzing patterns of distortion apart from the mean intensity and contrast of the image, thus providing a better approximation of the human visual system [159]. Our hypothesis is that the same approach can be effective in detecting bilateral asymmetry, through the quantification of structural similarity between paired mammographic regions. In the spatial domain, the SSIM index compares local patterns of pixel intensities that have been normalized for mean intensity and contrast [159]. Given a pair of right and left rectangular regions, $\mathbf{x}_R$ and $\mathbf{y}_L$, of size $M \times N$ and $P \times Q$ pixels, respectively, we introduce a Correlation-Based Structural SIMilarity (CB-SSIM) index to allow direct estimation of structural similarity between regions of different sizes, as follows:

$$S(\mathbf{x}_R, \mathbf{y}_L) =$$
$$= \frac{(2\mu_R \mu_L + K_1)\{2 \max\left[corr(\mathbf{x}_R, \mathbf{y}_L)\right] + K_2\}}{(\mu_R^2 + \mu_L^2 + K_1)\{\max\left[corr(\mathbf{x}_R, \mathbf{x}_R)\right] + \max\left[corr(\mathbf{y}_L, \mathbf{y}_L)\right]\}}, \quad (5.6)$$

where μ_R and μ_L are the mean values of pixels within the right and left regions, respectively, and

$$corr(\mathbf{x}_R, \mathbf{y}_L) =$$
$$\sum_{m=1}^{M} \sum_{n=1}^{N} \left[\mathbf{x}_R(m,n) - \mu_R\right]\left[\mathbf{y}_L(m-p, n-q) - \mu_L\right], \quad (5.7)$$

with $-P + 1 \le p \le M - 1$ and $-Q + 1 \le q \le N - 1$, is the cross-correlation in the 2D space between the right and left regions; $corr(\mathbf{x}_R, \mathbf{x}_R)$ and $corr(\mathbf{y}_L, \mathbf{y}_L)$ are the corresponding autocorrelation functions. K_1 and K_2 are two small positive constants to improve the robustness of the index: they were set to 0.01 and 0.03, respectively, as indicated in the work by Wang et al. [159]. Note that, if $N = M$ and $P = Q$, the index is equal to the SSIM index by Wang et al. [159] and that the value of unity is achieved if the regions $\mathbf{x}_R$ and $\mathbf{y}_L$ are identical. The concept of spatial structural similarity was extended to the complex wavelet domain by Sampat et al. [131]. The main advantage of CW-SSIM with respect to SSIM is insensitivity to scale and image distortions that are not related to the actual differences in the structure of the image. In particular, the use of steerable pyramid decomposition [117, 136] has been shown to be effective in the

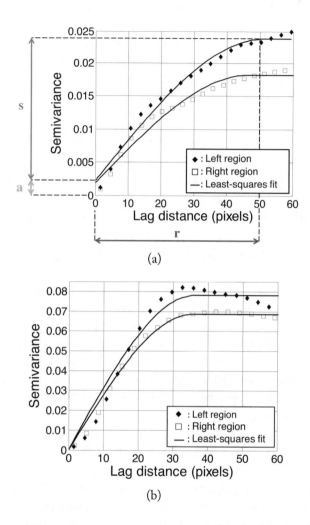

(a)

(b)

Figure 5.6: Empirical semivariograms and the corresponding spherical least-squares fits related to the left (black diamonds) and right (white squares) breast regions extracted from (a) the asymmetric malignant case A-1725-1 of DDSM [61] and (b) the normal case mdb035/mdb036 of mini-MIAS [141]. The nugget, a, the sill, s, and the range, r, are indicated. The corresponding differential spherical semivariogram descriptor values are: $\Delta V_a = 0.4 \times 10^{-3}$, $\Delta V_s = 4.2 \times 10^{-3}$, and $\Delta V_R = 4.2$ pixels, for the malignant asymmetric case, and $\Delta V_a = 0.2 \times 10^{-10}$, $\Delta V_s = 9.5 \times 10^{-3}$, and $\Delta V_R = 1.0$ pixels, for the normal case. Higher values of the descriptors indicate lower similarity between the paired regions analyzed and are expected to be correlated to the asymmetric cases. Reproduced with permission from Casti et al. [27] ©IEEE.

computation of CW-SSIM due to its rotation-invariance properties [131]. This aspect is crucial in applications for mammography, and especially in bilateral mammographic comparison, where the distortions caused by compression and relative translation of the two breasts during the imaging procedure may cause FP results. Therefore, similarly to what we did in the spatial domain, we define a Correlation-Based Complex Wavelet SIMilarity index, CB-CW-SSIM, as follows:

$$\widetilde{S}(c_R, c_L) = \frac{2 \max_{p,q} \left\{ | \sum_{i,j} \left[c_R(i,j) \, c_L^*(i-p, j-q) \right] | \right\}}{\sum_{i,j} |c_R(i,j)|^2 + \sum_{s,t} |c_L(s,t)|^2}, \tag{5.8}$$

where c_R and c_L are the complex wavelet coefficients obtained, respectively, by decomposing the regions x_R and y_L with a 3-scale, 16-orientation steerable pyramid decomposition procedure [117, 131, 136]. The asterisk denotes the complex conjugate.

Examples related to the computation of the correlation-based structural similarity indices, S and $\widetilde{S}$, are shown in Figs. 5.7 and 5.8 for the Gabor magnitude and phase responses of the central regions extracted from a malignant asymmetric case and a normal case. The relative scale between the left and right regions has been preserved to illustrate size differences related to corresponding paired areas of the two breasts of a patient. The corresponding feature values obtained for the two cases are also provided. Analysis of the cross-correlation functions shown in Figs. 5.7 and 5.8 and the related feature values indicate relatively low values of similarity for both cases. This is due to the inherent differences between the two breasts of a patient and to the additional dissimilarity introduced by compression and positioning of the breast during the mammographic examination. The normal case in Fig. 5.8, however, shows more diffuse areas of higher cross-correlation and, as expected, higher values of the corresponding structural similarity features than the asymmetric case in Fig. 5.7.

5.10.3 CLASSIFICATION OF MAMMOGRAMS AS ASYMMETRIC OR NORMAL PAIRS

The classification performance of each of the features extracted from the bilateral regions without training any classifier was first analyzed in terms of area under the ROC curve, A_z, for the two datasets of mammograms, DDSM and mini-MIAS, individually and combined. Stepwise logistic regression (SWR) [41] based on the F-statistic was used for automatic selection of the features in the training set of each experiment. Various p-values of the F-statistic were set to select different combinations of features and the obtained best results are reported in this work, together with the sets of the most frequently selected features. LDA, the BQDA classifier [42], and a two-layer artificial neural network with radial basis functions (ANN-RBF) [60] were used for classification. Each pair of mammograms was analyzed individually and, in addition, combination of the features extracted from the CC and MLO projections of the same patient was performed for the DDSM dataset to explore the performance of two-view analysis. The leave-one-patient-out method was used for cross-validation of results. Two-fold cross-validation was also applied to the whole set of mammograms, DDSM + MIAS, including 47 asymmetric

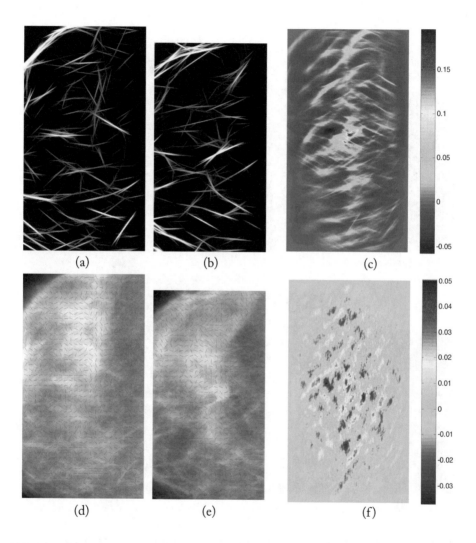

Figure 5.7: Analysis of cross-correlation of (a,b) Gabor magnitude and (d,e) phase responses for the central regions extracted from the asymmetric case A-1725-1 of DDSM [61]. (a,d) Left regions. (b,e) Right regions. The relative scale between the left and right regions has been preserved to illustrate size differences related to corresponding paired areas of the two breasts of the patient. (c,f) Normalized cross-correlation functions. The corresponding structural similarity descriptor values are: $S_M = 2.2 \times 10^{-13}$ and $S_\Phi = 3.2 \times 10^{-6}$ in the spatial domain, and $\widetilde{S}_M = 2.2 \times 10^{-6}$ and $\widetilde{S}_\Phi = 1.3 \times 10^{-6}$ in the complex wavelet domain, respectively, for the Gabor magnitude (M) and phase (Φ) responses. Reproduced with permission from Casti et al. [27] ©IEEE.

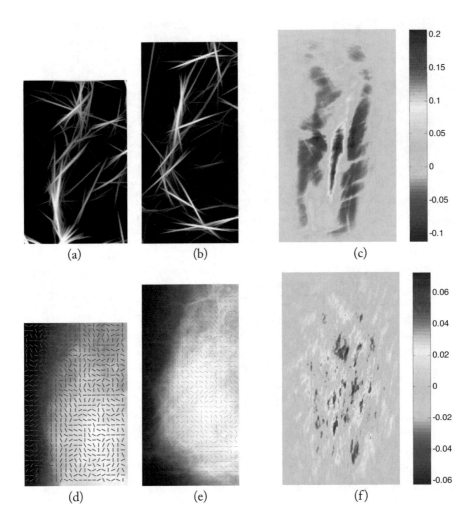

Figure 5.8: Analysis of cross-correlation of (a,b) Gabor magnitude and (d,e) phase responses for the central regions extracted from the normal case mdb035/mdb036 of mini-MIAS [141]. (a,d) Left regions. (b,e) Right regions. The relative scale between the left and right regions has been preserved to illustrate size differences related to corresponding paired areas of the two breasts of the patient. (c,f) Normalized cross-correlation functions. The corresponding structural similarity descriptor values are: $S_M = 3.9 \times 10^{-12}$ and $S_\Phi = 1.4 \times 10^{-4}$ in the spatial domain, and $\widetilde{S}_M = 7.1 \times 10^{-6}$ and $\widetilde{S}_\Phi = 7.1 \times 10^{-6}$ in the complex wavelet domain, respectively, for the Gabor magnitude (M) and phase (Φ) responses. Reproduced with permission from Casti et al. [27] ©IEEE.

and 47 normal cases, to test the robustness of the proposed approach; the procedure was repeated 100 times and the results averaged over the repetitions. The A_z (and standard error, SE) and 95% confidence interval, $I_{95\%}$, values were estimated using ROCKIT [140]. Sensitivity, specificity, and accuracy rates were computed at the operating point on the experimental ROC curve closest to the vertex $(0, 1)$.

5.10.4 RESULTS AND DISCUSSION

Performance of Individual Features

The classification performance of individual features in discriminating between normal and asymmetric pairs of mammograms is reported in terms of A_z in Table 5.5 for various datasets of images. Values higher than 0.5 indicate behavior according to expectation, i.e., in the case of the differential spherical semivariogram descriptors, ΔV, lower values for normal pairs and higher values for asymmetric pairs. The opposite trend is expected, instead, in the case of the similarity indices, S and $\widetilde{S}$. The obtained values of A_z indicate that all of the similarity indices follow the expected trend. The highest value of 0.88 was obtained by the CB-CW-SSIM index applied on the intensity values of the central regions of the two mammograms, $\widetilde{S}_I$, for the DDSM dataset. Analysis of the results indicates that CW-SSIM and CB-CW-SSIM possess stronger discriminating ability than the spherical semivariogram descriptors. The correlation-based similarity approach facilitates comparison between images that are inherently different in origin and size while sharing some degree of similarity that needs to be quantified.

   We compared the performance achieved by the proposed correlation-based descriptors with respect to the previously proposed structural similarity indices, SSIM [159] and CW-SSIM [131], applied to the various mammographic regions extracted as described in Sections 5.6 and 5.7. Since point-by-point correspondence is required for such analysis, regions of equal size were derived by removing the extra pixels along the x and y directions and then used for computation of the features. Table 5.6 summarizes the A_z values achieved by the SSIM and CW-SSIM indices in the classification of normal vs. asymmetric pairs of mammograms. The results indicate a general decrease in the performance of the individual features, SSIM and CW-SSIM, with respect to the performance achieved with our CB-SSIM and CB-CW-SSIM indices summarized in Table 5.5. Poorer results were observed, in particular, for the mini-MIAS dataset, for which a more effective matching between the compared regions appears to be important. The results also point out that the use of more sophisticated descriptors is needed for comparative analysis of the directional components of pairs of mammograms, as indicated by the lower A_z values obtained by the SSIM and CW-SSIM indices derived from the magnitude and phase responses of Gabor filters.

Performance of Classification and Cross-validation

The results of pattern classification using LDA, BQDA, and ANN-RBF classifiers with the leave-one-patient-out cross-validation method are reported in Table 5.7. The sets of features

Table 5.5: Performance of individual features for the classification of normal vs. asymmetric pairs of mammograms. Results are given in terms of A_z for various datasets of images. Cases with $A_z > 0.8$ are in bold.

Feature	Description	DDSM	MIAS	DDSM + MIAS
ΔV_{a1}	Nugget from the whole breast region	0.59	0.44	0.54
ΔV_{a2}	Nugget via masking #1	0.69	0.43	0.60
ΔV_{a3}	Nugget via masking #2	0.61	0.40	0.55
ΔV_{s1}	Differential sill from the whole breast region	0.51	0.40	0.47
ΔV_{s2}	Sill via masking #1	0.47	0.49	0.46
ΔV_{s3}	Sill via masking #2	0.55	0.35	0.51
ΔV_{r1}	Range from the whole breast region	0.57	0.56	0.56
ΔV_{r2}	Range via masking #1	0.65	0.49	0.60
ΔV_{r3}	Range via masking #2	0.64	0.72	0.65
ΔV_{g1}	Geometrical anisotropy from the whole breast region	0.68	0.65	0.67
ΔV_{g2}	Geometrical anisotropy via masking #2	0.54	0.29	0.47
ΔV_{g3}	Geometrical anisotropy via masking #3	0.47	0.48	0.47
S_{I1}	CB-SSIM from intensity values of the central region	0.63	0.65	0.59
S_{I2}	CB-SSIM from intensity values via masking #1	0.64	**0.80**	0.69
S_{I3}	CB-SSIM from intensity values via masking #2	0.61	0.65	0.57
S_{M1}	CB-SSIM from Gabor magnitude of the central region	0.75	0.72	0.72
S_{M2}	CB-SSIM from Gabor magnitude via masking #1	0.58	0.75	0.63
S_{M3}	CB-SSIM from Gabor magnitude via masking #2	0.53	0.65	0.57
$S_{\Phi1}$	CB-SSIM from Gabor phase of the central region	**0.85**	**0.82**	**0.84**
$S_{\Phi2}$	CB-SSIM from Gabor phase via masking #1	0.70	0.78	0.72
$S_{\Phi3}$	CB-SSIM from Gabor phase via masking #2	0.79	0.77	0.79
$\widetilde{S}_I$	CB-CW-SSIM from intensity values of the central region	**0.88**	0.80	**0.85**
$\widetilde{S}_M$	CB-CW-SSIM from Gabor magnitude of the central region	**0.81**	0.79	**0.81**
$\widetilde{S}_\Phi$	CB-CW-SSIM from Gabor phase of the central region	**0.84**	0.80	**0.81**

ΔV: *differential semivariogram descriptors. S and $\widetilde{S}$: structural similarity descriptors.*
CB-SSIM: Correlation-Based Structural SIMilarity index.
CB-CW-SSIM: Correlation-Based Complex Wavelet Structural SIMilarity index.
Masking #1: horizontal/annular masking. Masking #2: vertical/oblique masking.

Table 5.6: Performance of the SSIM and CW-SSIM indices for the classification of normal vs. asymmetric pairs of mammograms. Results are given in terms of A_z for the various datasets of images. Cases with $A_z > 0.8$ are in bold.

Description	DDSM	MIAS	DDSM + MIAS
SSIM from intensity values of the central region	0.71	0.55	0.62
SSIM from intensity values via masking #1	0.73	0.38	0.60
SSIM from intensity values via masking #2	0.66	0.56	0.59
SSIM from Gabor magnitude of the central region	0.51	0.58	0.53
SSIM from Gabor magnitude via masking #1	0.53	0.67	0.58
SSIM from Gabor magnitude via masking #2	0.57	0.48	0.55
SSIM from Gabor phase of the central region	0.71	0.65	0.69
SSIM from Gabor phase via masking #1	0.72	0.64	0.70
SSIM from Gabor phase via masking #2	0.65	0.55	0.60
CW-SSIM from intensity values of the central region	**0.81**	0.63	0.75
CW-SSIM from Gabor magnitude of the central region	0.63	0.57	0.60
CW-SSIM from Gabor phase of the central region	0.56	0.62	0.58
SSIM: Structural SIMilarity index.			
CW-SSIM: Complex Wavelet Structural SIMilarity index.			

selected more than 50% of the time in the training step via SWR are listed for each experiment. The best A_z (SE) values obtained on a per-pair-of-mammogram basis for the DDSM and MIAS datasets individually were, respectively, 0.90 (0.04) and 0.88 (0.04). Analysis of the corresponding ROC curves indicated accuracies up to 0.91 and 0.87, respectively, for the DDSM and MIAS datasets. The combination of the two datasets on a per-pair-of-mammograms basis, DDSM + MIAS, led to A_z of 0.83 (0.04), 0.77 (0.05), and 0.87 (0.04), respectively, with the LDA, BQDA, and ANN-RBF classifiers. The best accuracy achieved was with the BQDA classifier when two of the similarity indices were selected often.

Only for the DDSM dataset, for which both CC and MLO views were available, two-view analysis was performed by combining the features extracted from the CC and MLO views on a per-patient basis. The overall best performance was achieved with two-view analysis of the DDSM dataset with the ANN-RBF classifier, with the A_z value of 0.93 (0.04), with the corresponding sensitivity, specificity, and accuracy of 1, 0.88, and 0.94, respectively, on a per-patient basis. The binormal ROC curves estimated by ROCKIT related to two-view analysis are displayed in Fig. 5.9 for the three classifiers used. Results for the normal pair vs. asymmetric pair classification for the 94 pairs of mammograms of the combined dataset (DDSM + MIAS) using the features selected via SWR and the LDA classifier for several cross-validation methods

Table 5.7: Results of pattern classification using the features selected via stepwise logistic regression and the leave-one-patient-out cross-validation procedure. Results are provided on a per-pair-of-mammograms basis (single view) for the various datasets of images and on a per-patient basis (two-view) for the DDSM. Features selected more than the 50% of the time during the cross-validation process are listed. Sensitivity, specificity, and accuracy rates were computed at the operating point on the experimental ROC curve closest to the vertex $(0, 1)$.

Database	Classifier	Selected Features	A_z (SE)	$I_{95\%}$	Sen.	Spec.	Acc.
DDSM (single-view)	LDA	$S_{M3}, \widetilde{S}_{\mathrm{I}}$	0.90 (0.04)	[0.79 - 0.95]	0.81	0.91	0.86
	Bayesian	$S_{M3}, \widetilde{S}_{\mathrm{I}}$	0.83 (0.05)	[0.72 - 0.91]	0.84	0.84	0.84
	ANN-RBF	$\Delta V_{a2}, S_{M3}, \widetilde{S}_{\mathrm{I}}$	0.90 (0.04)	[0.78 - 0.96]	0.84	0.97	0.91
MIAS (single-view)	LDA	$\Delta V_{a2}, \Delta V_{r3}, \Delta V_{g2}, \widetilde{S}_{\Phi}$	0.87 (0.07)	[0.69 - 0.96]	0.80	0.93	0.87
	Bayesian	$\Delta V_{a2}, \Delta V_{s1}, \Delta V_{s3}, \Delta V_{r3},$ $\Delta V_{g2}, S_{I1}, S_{\Phi1}, S_{\Phi2}, \widetilde{S}_{\Phi}$	0.84 (0.08)	[0.64 - 0.95]	0.80	0.93	0.87
	ANN-RBF	$\widetilde{S}_{\Phi}$	0.88 (0.06)	[0.71 - 0.96]	0.87	0.80	0.83
DDSM + MIAS (single-view)	LDA	$\widetilde{S}_{\mathrm{I}}$	0.83 (0.04)	[0.73 - 0.90]	0.68	0.89	0.79
	Bayesian	$S_{M3}, \widetilde{S}_{\mathrm{I}}$	0.77 (0.05)	[0.65 - 0.85]	0.84	0.80	0.83
	ANN-RBF	$S_{M3}, \widetilde{S}_{\mathrm{I}}$	0.87 (0.04)	[0.79 - 0.93]	0.74	0.87	0.81
DDSM (two-view)	LDA	CC: $\Delta V_{s1}, \Delta V_{r2}, \Delta V_{g1}$ MLO: $\Delta V_{s1}, \widetilde{S}_{\mathrm{I}}$	0.85 (0.07)	[0.68 - 0.95]	0.81	0.75	0.78
	Bayesian	CC: $\Delta V_{s1}, \Delta V_{r2}, \Delta V_{g1}$ MLO: $\Delta V_{s1}, \widetilde{S}_{\mathrm{I}}$	0.78 (0.09)	[0.58 - 0.91]	0.75	0.75	0.75
	ANN-RBF	CC: $\Delta V_{s2}, \Delta V_{r2}, \Delta V_{g1}, S_{M3}$ MLO: $\Delta V_{s1}, \widetilde{S}_{\mathrm{I}}$	0.93 (0.06)	[0.73 - 0.99]	1	0.88	0.94

are summarized in Table 5.8. As expected, higher values of A_z, up to 0.86, were obtained when the features were selected using the entire dataset of mammograms.

Discussion

Detection of asymmetric signs of breast disease in mammograms is an important but challenging problem. In this study, analysis of structural similarity was performed via landmarking, automatic bilateral masking procedures, multidirectional Gabor filtering, modeling of spherical semivariograms, and extraction of similarity features. The last step was performed by introducing correlation-based structural similarity indices in both spatial (CB-SSIM) and complex wavelet (CB-CW-SSIM) domains, which extended the SSIM and CW-SSIM indices to facili-

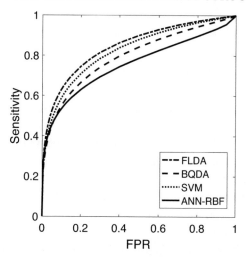

Figure 5.9: Binormal ROC curves estimated using ROCKIT [140] for the dataset of 32 cases of the DDSM (two-view analysis) using leave-one-patient-out cross-validation and SWR for feature selection. The curves represent the performance of classification of normal vs. asymmetric cases on a per-patient basis. Reproduced with permission from Casti et al. [27] ©IEEE.

Table 5.8: Results of ROC analysis for the 94 cases of the combined dataset of mammograms (DDSM + MIAS) using the features selected via SWR and the LDA classifier for several cross-validation methods

Cross-validation Method	Feature Selection Using the Training Set	Featue Selection Using the Entire Dataset
LOO, pair of images	0.83	0.86
LOO, patient	0.83	0.85
2-fold, pair of images	0.81 ± 0.03	0.86 ± 0.02
2-fold, patient	0.80 ± 0.04	0.82 ± 0.03

tate quantitative comparison of regions of different sizes for which point-by-point comparisons are inherently difficult and not meaningful.

Appropriate matching between corresponding regions of the left and right mammograms was performed by means of automatically detected landmarks, i.e., the nipple, the breast boundary, and the pectoral muscle (only for MLO views), which were used as geometrical references for performing bilateral masking procedures. The accuracy in the detection of such landmarks may influence the classification performance of some of the individual features and, consequently, of the whole CAD system. Effective landmarking is important for the proposed CAD system to derive anatomical correspondence between the left and right mammographic regions to be compared. However, relatively small inaccuracies in the detection of the reference anatomical structures are coped with by avoiding point-by-point comparisons. In addition, registration procedures that may alter the structure of the breast parenchyma based on the extracted landmarks have not been performed. At the same time, the correlation-based structural similarity approach facilitates comparison between images of different size and origin, for which forcing perfect matching would be inherently inappropriate. The strength of our approach was demonstrated in Section 5.10.4 by comparing the performance of the proposed correlation-based similarity descriptors with the performance achieved using the point-by-point methods conventionally used for image quality assessment.

Analysis of the performance of individual features proved the efficacy of the extracted features in detecting asymmetric signs in the distribution of the fibroglandular components of the two mammograms of a patient. CB-CW-SSIM, in particular, showed the best discriminatory power, especially for the case of MLO views of the MIAS dataset, where the differential distortions due to positioning and compression of the breast are more evident. Results of pattern classification showed that the features also complement one another and can be combined via automated feature selection procedures to achieve higher performance levels. The variability in the various sets of features selected summarized in Table 5.7 indicates that the information content in the training data differs among the different combinations of datasets used for the various pattern classification experiments. Additional studies with larger sets of mammograms will be helpful to determine the optimal set of features and the most effective pattern classification model.

The major limitation of the present study is the use of a relatively small dataset of mammograms for development and validation of the proposed methods. However, all of the asymmetric cases available in public databases of mammograms were included in the analysis presented in this work. More extensive testing of the proposed procedures on larger datasets would be desirable to confirm our results and proceed toward clinical application.

Comparative analysis with other reports on detection of bilateral asymmetry in mammograms [48, 81, 98, 124, 150, 156, 157] (see Table 5.1) indicates that the results obtained in this study are better than the results in previous studies, including our previous work [21, 25, 93]. An A_z value of 0.78 was recently reported by Wang et al. [157] combining statistical, textural, and density features. Their method was tested on a private dataset of mammograms by means of an ANN classifier and the leave-one-patient-out cross-validation procedure. The results obtained by other studies on the mini-MIAS database [141] with leave-one-patient-out cross-validation were reported in terms of accuracy [48, 124, 150]. Among them, the methods based on directional analysis by means of Gabor wavelets and rose diagrams achieved accuracy up to 0.84 [48, 124]. The work based on differential first-order statistical features led to an accuracy of 0.86 [150]. However, the last mentioned study did not consider only the asymmetric cases of the mini-MIAS for the analysis, but also masses, architectural distortion, and calcifications.

In the present study, which is entirely focused on the detection of bilateral asymmetry by using all of the asymmetric cases available in public databases of mammograms, we have demonstrated the ability of the proposed techniques to detect pathological differences in the fibroglandular components of breast tissue patterns in mammograms with accuracy up to 0.94. An accuracy of 0.87, with the corresponding sensitivity and specificity of 0.80 and 0.93, respectively, was achieved using all of the asymmetric cases of the mini-MIAS database. The results obtained with the DDSM cases also point out that the combination of the information extracted from CC and MLO views can provide higher accuracy than the single-view approach. The robustness of our approach is indicated by the good results obtained by combining mammograms from two different databases and acquired at multiple hospitals. The efficacy of the analysis of structural similarity has been furtherly investigated using ensembles of classification models based on artificial immune system [86, 87] and the masking procedures have been succesfully applied for localization of malignant sites of asymmetry on mammograms [28].

5.11 REMARKS

The importance of detecting bilateral asymmetry in mammograms has been discussed in this chapter and the related difficulties encountered by radiologists analyzed. Previous work dealing with automatic detection of asymmetric findings between the left and right breasts of a patient have been described, stressing the need of new algorithmic solutions. New methods for detection of bilateral asymmetry based upon the analysis of phase and structural similarity and of spatial correlation have been presented, which make use of Tabár masking procedures for bilateral comparison of paired mammographic regions. The reported results have demonstrated the effectiveness of the proposed strategies to detect bilateral structural asymmetry. The related limitations have been also discussed. Performance analysis of the three methods indicates that the automated masking procedures can be an aid in improving the results of pattern classification based on comparison of the directional structures of the two breasts, as long as accurate matching between bilateral pairs of corresponding regions is achieved.

The following chapters describe automatic procedures for detection and diagnosis of other mammographic signs of breast cancer, including masses and architectural distortion.

CHAPTER 6

Design of Contour-independent Features for Classification of Masses

6.1 MOTIVATION

Assessment of the breast anomalies detected during mammography examinations is fundamental to define appropriate follow-up for patients. Incorrect interpretation of mammographic lesions by radiologists may lead to misdiagnosis of breast cancer. Progression of cancer and delays in needed treatment can result if malignant lesions are misinterpreted as being benign (FP diagnosis). Also, unwarranted biopsies and anxiety for the patient occur if benign masses are assessed as being malignant (FP diagnosis). Lack of experience, fatigue, and inattention of the radiologist, as well as the subtlety or indistinct nature of features of malignancy in masses as they appear on the mammogram are well-known causes of interpretation errors [88]. CADx can serve as an aid to radiologists in the classification of breast abnormalities in mammograms as benign lesions or malignant tumors.

6.2 STATE OF THE ART

Automatic classification of lesions starts from extracted ROIs or contours of the lesions delineated by radiologists on the mammogram [43]. Most of the existing approaches are based on accurate estimation of the contour of masses, either drawn by a radiologist or extracted by dedicated segmentation procedures [38, 43, 58, 102, 119, 122, 129]. Then, relevant features of masses are quantified in terms of their texture and shape in order to assess the likelihood of malignancy [43].

Mudigonda et al. [102] proposed a method for analyzing oriented textural information in adaptive ribbons of pixels across the margins of masses. Using a dataset of 32 automatically detected and segmented masses, an area under the ROC curve (A_z) of 0.79 was reported for the discrimination of benign and malignant lesions. Sahiner et al. [129] developed a three-stage method for accurate segmentation of masses to extract morphological and texture features. An A_z value of 0.91 ± 0.02 was obtained using 249 SFMs with leave-one-patient-out cross-validation. The study by Rangayyan and Nguyen [122] explored the use of fractal analysis to characterize masses based on shape. Although the results were promising ($A_z = 0.93$

with 111 breast masses), manual delineation of detailed contours of the masses by an expert radiologist was necessary.

In the work by Domínguez and Nandi [38], the performance of a set of six features, designed for characterization of margins of masses, was compared by varying the method used for segmentation. Contours obtained by a dynamic-programming-based method, a region growing algorithm, and by means of elliptical approximations of the initially obtained contours were used to predict the diagnosis of 349 masses. The authors indicated that the accuracy of the segmentation of masses influences the classification efficiency and recommended the design of new features that are robust to inaccuracies in the extracted boundaries. Hapfelmeier and Horsch [58] used a segmentation approach called concentric density regions to segment 1934 ROIs. Based on the mass segmentation result, 242 features were extracted, which led to $A_z = 0.86$. A semi-automatic procedure was proposed by Ramos-Pollán et al. [119], where the user optimized the automatically segmented boundaries manually. They reported $A_z = 0.996$ by combining feature vectors extracted from CC and MLO views of the same case.

In their review on CADx of mammographic masses, Elter and Horsch [43] emphasized the inherent difficulties in the segmentation of masses by questioning if a fully automatic and robust approach for the diagnosis of masses could be developed, and called for novel or semiautomatic solutions. In fact, if based on accurate estimation of the contours of the lesions, either drawn by a radiologist or extracted by dedicated segmentation procedures, the systems are prone to fail in the presence of obscured or ill-defined boundaries of tumors. The problem is made more difficult by the superimposition of the surrounding parenchyma on the margins of masses in radiographically dense breasts. Intraexpert and interexpert variability in identifying or validating the margins of lesions may raise questions concerning the reliability of such systems [43].

6.3   OVERVIEW OF THE DESIGN STUDIES

The aim of the following two studies is to overcome the limitations of the existing approaches to the classification of masses and to develop and test new features that do not strictly depend on the contours obtained and are not influenced by the accuracy with which the contours depict the masses. With this purpose, a 2D circular domain representation, or circular region of interest (C-ROI), including a mass is characterized by quantification of the nonstationarity and spatial dependence of pixel values.

Statistical techniques are conventionally used to determine whether certain characteristics in terms of space and time are present, with an associated confidence, in a given signal. Khademi et al. [71] applied a two-dimensional (2D) extension of the reverse arrangement (RA) test [6] and Mantel's test for clustering [90] to examine, respectively, nonstationarity and spatial dependence of pixel values in a set of phantom images. They also explored the possibility of applying such tests to generate feature descriptors. Such possibility is addressed in our Study 1 by the development of features for the analysis of correlation and trend in the radial direction [20], while in Study 2, the same feature descriptors are extended to the angular direction [21]. The angular measures designed in Study 2 are made invariant to rotation via the computation of the principal axis of the grayscale values of the mass in a C-ROI. The required inputs to extract the C-ROIs are the centroid of the lesion and the maximum radial distance of the centroid from the contour manually drawn by the radiologist (Study 1) or an average radius and a randomly selected centroid within the mass contour (Study 2). Performance analysis and results of cross-validation are reported to verify the applicability of the proposed features in the absence of a contour. In Chapter 7, the designed features will be applied for the task of diagnosis based on automatically detected C-ROIs.

6.4   STUDY 1: DESIGN AND PERFORMANCE ANALYSIS OF RADIAL FEATURES

6.4.1   EXPERIMENTAL SETUP

A total of 146 ROIs, including 120 benign masses and 26 malignant tumors, were extracted from the FFDM images (see Section 2.1.1) using the contours manually drawn by expert radiologists specialized in mammography for this design study. The centroid of each mass was computed by using the first-order moments of the grayscale values within its contour weighted by their distances from the center of the axes. Then, each ROI was automatically sized in order to include the manually drawn contour and centered at the centroid of the mass. Results of biopsy providing the diagnostic classification of each mass were used to validate the proposed approach. Figures 6.1a and e show two examples of the ROIs used in this work, including, respectively, a benign mass and a malignant tumor.

6.4.2   METHODS

Preprocessing Stage

The original ROIs were downsampled to the spatial resolution of 200 for reducing the computational cost of the proposed methodology. Then, in order to enhance properly the contrast of the masses, the look-up table (LUT) information of the softer linear transformation encoded in the DICOM tags [103] of the FFDMs was applied to the grayscale values of the ROIs. In Figs. 6.1b and f, two examples of ROIs after the LUT transformation are shown, including, respectively, a benign mass and a malignant tumor. The contours drawn by the expert radiologist and the centroids of the masses are superimposed on the ROIs.

A C-ROI of radius R centered at the centroid of the lesion was extracted from each ROI (see Figs. 6.1c and g). Note that the two ROIs in Fig. 6.1 have been resized to the same dimension for display purposes. As a first attempt, for each ROI, the maximum radial length (R_m), computed as the maximum of the Euclidean distances from the centroid of the mass to each of the boundary coordinates, is used as the radius of the C-ROI.

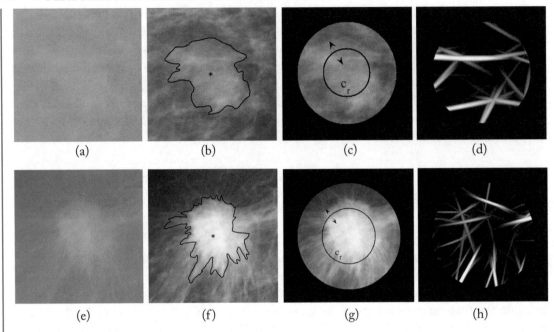

Figure 6.1: Original ROI including (a) a benign mass and (e) a malignant tumor. (b,f) ROI after the LUT transformation. The contour manually drawn by the expert radiologist and the computed centroid are superimposed on the image. (c,g) Circular domain centered at the centroid of the mass. A ring c_r of a given radius R, (c) $R = 40$ pixels and (g) $R = 80$ pixels, is superimposed on the image. (d,h) Gabor magnitude response. Note that the two ROIs were resized to the same dimension for display purposes. Reproduced with permission from Casti et al. [20] ©IEEE.

Extraction of Oriented Patterns

In order to quantify trend and correlation of the directional components present within a mass, a set of 18 real Gabor filters [53] equally spaced in the angular range $(-\pi/2, \pi/2]$ was applied by means of the procedure described in Section 3.1. Examples of the magnitude response used for the extraction of features in addition to the intensity values are shown in Figs. 6.1d and h. A Gabor kernel with $\tau = 24$ pixels and $l = 4$ was used in this study.

Radial Correlation Measure

The first feature was designed to measure radial correlation within a given C-ROI. Let us virtually divide the C-ROI into $R - 1$ annuli of radius r_i, $i = 1, 2, \ldots, R - 1$ pixels, as shown in Fig. 6.2a, and consider the average of the pixel values μ_i^r, $i = 1, 2, \ldots, R - 1$, of each annulus. Note that μ_1 is the pixel value at the centroid of the mass. A generic ring of radius r is superimposed on the image in each case in Figs. 6.1c and g.

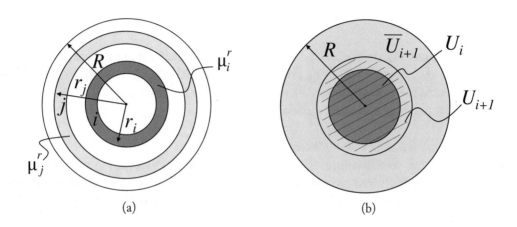

(a)   (b)

Figure 6.2: Schematic representation of the computation of (a) the measure of radial correlation and (b) the measure of radial trend for a C-ROI of radius R. U_i indicates the ith annulus and $\overline{U}_i$ the corresponding complement in the circular domain [20, 23]. Note: $U_i \in U_{i+1}$. Reproduced with permission from Casti et al. [26] ©Elsevier.

If radial correlation or radial dependence exist among the pixels in the C-ROI, annuli close to one another will be coupled with average intensity values which are also similar. In order to quantify the extent of correlation in the radial direction, a radial correlation measure, C_R, based on Mantel's test for clustering, is defined as

$$C_R = \frac{1}{R_{tot}} \sum_{i=1}^{R} \sum_{j=1}^{R} |r_i - r_j||\mu_i^r - \mu_j^r|, \qquad (6.1)$$

where $R_{tot} = \sum_{i=1}^{R} \sum_{j=1}^{R} |r_i - r_j|$ is the sum of all the radial separation values over the C-ROI. Note that a homogeneous distribution of pixel values in the C-ROI would result in $C_R = 0$. Higher values of C_R are indicative of lower spatial dependency or heterogeneity among pixels in the radial direction, which is typical of malignant tumors, whereas lower values of C_R indicate radial clustering, which is typical of benign masses.

Radial Trend Measure
Let us consider the region U_1 representing the central pixel of the C-ROI. Starting from U_1, a total of $R - 1$ concentric circles of increasing area are defined as shown in Fig. 6.2b, and the averages of the pixel values M_i^r, $i = 1, 2, \ldots, R - 1$, within these regions are computed. A radial trend measure based on a novel 2D extension of the RA test is proposed, which involves

computing the number of times that pixel values within the complement of U_i, $\overline{U}_i$, in the C-ROI are less than M_i^r, as

$$T_R = \frac{1}{\#\overline{U}_{tot}} \sum_{i=1}^{R} \#\overline{U}_i < M_i^r.$$

(6.2)

The reverse arrangements are normalized by $\#\overline{U}_{tot}$, which is the total number of pixels within the complement $\overline{U}_i$. Higher values of T_R are indicative of radial trend (non-stationarity), which can be correlated to the characteristics of a malignant tumor infiltrating the surrounding tissue in the radial direction (e.g., radiating spicules).

Pattern Classification and Cross-validation

To avoid bias, feature selection using SWR [120] and pattern classification were performed using the leave-one-patient-out cross-validation method. All of the four features were selected more than 50% of the time. Four classifiers, FLDA, BQDA, SVM [42], and ANN-RBF [60] were used.

6.4.3   RESULTS AND DISCUSSION

The two proposed measures of radial trend and radial correlation described above were computed for all of the ROIs obtained from the FFDM database of mammograms. For each ROI, four features were extracted by computing the two measures from the original ROI after the LUT transformation (T_{RI}, C_{RI}) and the Gabor magnitude response (T_{RM}, C_{RM}).

The performance of each feature was, at first, analyzed independently without training any classifier by means of the ROCKIT package [140]. The obtained individual A_z values are listed in Table 6.1. The A_z values indicate satisfactory to good performance in the classification of benign masses vs. malignant tumors.

Table 6.1: List of features for characterization of radial correlation and radial trend with A_z values

Symbol	Feature	A_z
C_{RI}	Original ROI after LUT transformation	0.83
C_{RM}	Gabor magnitude response	0.83
T_{RI}	Original ROI after LUT transformation	0.82
T_{RM}	Gabor magnitude response	0.77
C_{RI}, C_{RM}: radial correlation measures; T_{RI}, T_{RM}: radial trend measures. The Az values were estimated using ROCKIT.		

In order to evaluate the dependence of the performance of the features on the dimension of the C-ROI, the A_z value was computed, for each feature, by varying the radius of the domain in the range $[1/2\, R_m, R_m]$, where R_m is the maximum radial length defined in Section 6.4.2.

Figure 6.3 illustrates the effect of varying the radius on the A_z values. Experimental results show that the proposed features are robust to variations in the dimension of the C-ROI. Values of A_z

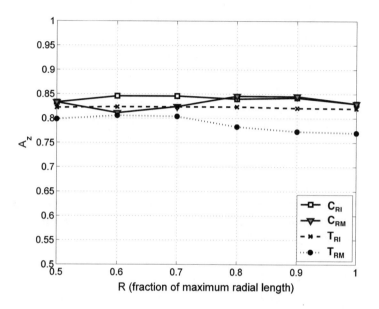

Figure 6.3: Effect of varying the radius of the C-ROI on the A_z values of the proposed features. C_{RI}, C_{RM}: radial correlation measures; T_{RI}, T_{RM}: radial trend measures. Reproduced with permission from Casti et al. [20] ©IEEE.

(and standard error, SE) of 0.82 (0.05), 0.76 (0.06), 0.85 (0.05), and 0.93 (0.03) were obtained, respectively, with the four classifiers listed above. Table 6.2 lists the obtained results along with the asymmetric 95% confidence intervals estimated using ROCKIT. The high values of A_z obtained, especially 0.93 with the ANN-RBF, indicate that the four features complement one another and can be used together to achieve increased classification efficiency. The performance achieved is comparable to the best results reported in the literature [122, 123, 128, 129]. In particular, in the work by [129], $A_z = 0.91$ is reported when the leave-one-patient-out method was applied to partition the dataset into the training and test sets. The estimated binormal ROC curves are shown in Fig. 6.4.

The obtained results show that the features proposed in the present study could be used to assess the likelihood of malignancy of breast masses in mammograms. The main advantage of the proposed methodology is that its effectiveness is independent of the accuracy with which the contours used depict the masses. In fact, the quantification of radial correlation and radial trend among pixels only requires a circular region properly centered on the lesion. This is confirmed by the low variability in the classification performance of the features in relation to the size of the C-ROI used in their computation. Although the ROIs in the present work were initially obtained

Table 6.2: Results of ROC analysis (A_z) using different types of classifiers

Classifier	A_z	SE	$I_{95\%}$
FLDA	0.82	0.05	[0.72, 0.90]
BQDA	0.76	0.06	[0.63, 0.87]
SVM	0.85	0.05	[0.74, 0.92]
ANN-RBF	0.93	0.03	[0.85, 0.98]

FLDA: Fisher linear discriminant analysis; BQDA: Bayesian classifier; SVM: support vector machine; ANN-RBF: articial neural network classifier based on radial basis functions. The Az values were estimated using ROCKIT along with the related standard error (SE) and the asymmetric 95% confidence interval (I95%).

with reference to the contours of the masses drawn by a radiologist, this step is not required in the proposed procedure since only the C-ROI is used for computation of the features. The centroid and a range of the expected size of the mass are required to define the C-ROI used to compute the features. However, the centroid of the mass could be detected by means of automated procedures or manually pointed by a radiologist. The range of the size of masses expected in a certain dataset or population could be estimated from previous data or experience.

6.5   STUDY 2: DESIGN AND PERFORMANCE ANALYSIS OF ANGULAR FEATURES

6.5.1   EXPERIMENTAL SETUP

The dataset of ROIs including the 146 masses was also used in this study. As described in Section 6.4.1, the manually drawn contour was used to extract the centroid and the maximum radial length of each mass, with the latter defined as the distance from the centroid to the farthest pixel on the contour. A radius, R, was assigned to each mass in a leave-one-patient-out procedure, using a measure of the expected radius of the C-ROI estimated by averaging the maximum radial lengths of the remaining masses. Then, given the expected radius R of each mass, 10 displaced centroidal points were randomly selected in a square region of size $0.5\,R$ centered at the actual centroid, so that 10 more C-ROIs of radius R were extracted. The randomly selected centroidal points were used to analyze the robustness of the proposed features to errors in the detection of the center of the mass. Figures 7.2a and b show, respectively, an example of a benign mass and a malignant tumor, where the boundary of the C-ROI centered at the centroid of the lesion is superimposed on each image (black solid line), together with three of the 10 C-ROIs centered at the randomly selected centroidal points.

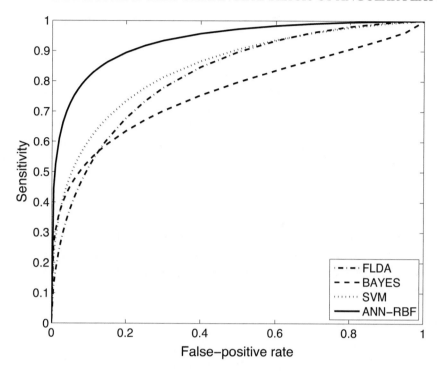

Figure 6.4: Estimated binormal ROC curves using ROCKIT for the dataset of 146 ROIs of masses using the selected features based on SWR and four different types of classifiers with the leave-one-patient-out method. The classifiers used are Fisher linear discriminant analysis (FLDA), Bayesian classifier (BQDA), support vector machine (SVM), and artificial neural network classifier based on radial basis functions (ANN-RBF).

6.5.2   METHODS

Computation of the Principal Axis

The principal axis [127] of the gray-level values within each C-ROI was used as the reference orientation for the computation of the angular features and to make them invariant with respect to rotation. Two principal axes are defined in the 2D space by the slopes, $\Theta_{1,2}$, that minimize the moment of inertia of the intensity values, $f(x, y)$, about the line, $y - \bar{y} = (x - \bar{x}) \tan \theta$, passing through the centroid with the coordinates $(\bar{x}, \bar{y})$:

$$\Theta_{1,2} = \operatorname*{argmin}_{\theta} \sum_{u,v} [(x - \bar{x}) \sin \theta - (y - \bar{y}) \cos \theta]^2 \, f(x, y). \tag{6.3}$$

The two solutions, $\Theta_{1,2}$, can be obtained by means of the matrix of the second-order central moments:

$$\bar{M} = \begin{pmatrix} \bar{m}_{20} & \bar{m}_{11} \\ \bar{m}_{11} & \bar{m}_{02} \end{pmatrix},$$

(6.4)

where $\bar{m}_{ij} = \sum_{x,y} (x - \bar{x})^i (y - \bar{y})^j f(u, v)$ is the central moment of $f(x, y)$ of order i, j. The angle of the principal axis, Θ^*, is given by the direction of the eigenvector corresponding to the smaller eigenvalue of $\bar{M}$ [127]. Figures 6.5c and d illustrate the results obtained for a benign case and a malignant case, respectively. The C-ROI extracted from the actual centroid is shown in the upper-left part of each of the two figures, while the remaining C-ROIs were extracted using the randomly selected centroidal points.

Extraction of Oriented Patterns

The extraction of oriented patterns was performed using a set of 18 real Gabor filters [53], with $\tau = 24$ pixels and $l = 4$, equally spaced over the angular range $(-\pi/2, \pi/2]$ (see Section 3.1 for more details on analysis of directional components). Figures 6.5e and f show the magnitude response obtained for two examples of ROIs.

Angular Correlation Measure

Let us consider a C-ROI of radius R and divide it into N equally spaced sectors ($N = 36$ in the present work), so that the central angle of one of them is aligned with the principal axis Θ^*, as shown in Fig. 6.6a. Using the average of the pixel values μ_i^a, $i = 1, 2, \ldots, N$, of each sector, an angular correlation measure, C_A, can be defined as follows:

$$C_A = \frac{2}{N_{tot}} \sum_{i=1}^{N} \sum_{j=1}^{N} \Delta ij \, |\mu_i^a - \mu_j^a|,$$

(6.5)

where Δij (see Fig. 6.6a) is defined as the separation between the ith and the jth sectors:

$$\Delta ij = \begin{cases} |i - j| & \text{if } |i - j| \leq N/2, \\ N - |i - j| & \text{otherwise,} \end{cases}$$

(6.6)

and $N_{tot} = \sum_{i=1}^{N} \sum_{j=1}^{N} \Delta ij$ is the sum of all the angular separation values over the C-ROI. When the sectors have similar average intensity values, C_A is lower, whereas higher values of C_A indicate heterogeneity among pixels in the angular direction. The grayscale and oriented patterns (if present) of benign lesions are more likely to be uniform or homogeneous, resulting in lower values of C_A, which is indicative of angular clustering. On the contrary, the nonuniform texture typical of malignant tumors is expected to result in higher values of C_A.

Angular Trend Measure

Let us now consider the sector S_1 of the C-ROI centered at the orientation of the principal axis Φ (see Fig. 6.6b). Starting from S_1, a total of $N/2 + 1$ sectors of increasing area are defined

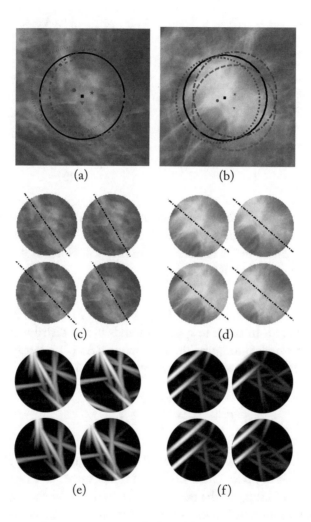

Figure 6.5: (a,b) Portions of mammograms including (a) a benign mass and (b) a malignant tumor after the LUT transformation. The boundary of the C-ROI centered at the centroid of the mass is superimposed on the image with a solid black line. The contours of three of the 10 C-ROIs centered at the randomly selected centroidal points are also shown. The estimated radius is $R = 52$ pixels (10.4 mm) for both cases. (c,d) C-ROIs of radius R centered at the centroid of the mass (upper left) and at the randomly selected centroidal points. The principal axis is superimposed on each C-ROI. (e,f) Gabor magnitude response restricted to the C-ROIs. Reproduced with permission from Casti et al. [21] ©IEEE.

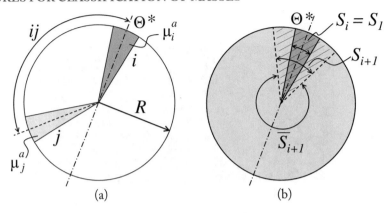

Figure 6.6: Schematic representation of the computation of (a) the measure of angular correlation and (b) the measure of angular trend for a C-ROI of radius R. S_i indicates the ith sector and $\overline{S}_i$ the corresponding complement in the circular domain [21, 23]. Note: $S_i \in S_{i+1}$. Reproduced with permission from Casti et al. [21] ©IEEE.

as shown in Fig. 6.6b, and the averages of the pixel values M_i, $i = 1, 2, \ldots, N/2 + 1$, within those regions are computed. In order to quantify trends in the distribution of pixel values that are indicative of angular nonstationarity, the number of RAs [7] is computed by summing the number of times that pixel values within the complement of S_i in the C-ROI are less than M_i, as

$$T_A = \frac{1}{\#X} \sum_{i=1}^{N/2+1} \#\overline{S}_i < M_i, \tag{6.7}$$

where $\overline{S}_i$ is the complement of S_i in the C-ROI. The total number of RAs is normalized by $\#X$, which is the number of pixels in the C-ROI. Higher values of T_A are indicative of greater angular nonstationarity and suggest the presence of large trends in the angular direction, which are expected in malignant tumors. On the contrary, stationarity of pixel values, resulting in lower values of T_A, is indicative of a benign mass.

Pattern Classification and Cross-validation

Quantification of the heterogeneity and nonstationarity between the sectors of the C-ROIs was performed by applying the proposed procedure to the gray-scale values (to obtain measures of angular correlation and trend of intensity, C_{AI} and T_{AI}) and to the Gabor magnitude response values (to obtain measures of angular correlation and trend of Gabor magnitude, C_{AG} and T_{AG}). Performance analysis of the features was carried out using A_z. The A_z (and standard error, SE) values were obtained using the ROCKIT package [140]. The classification of masses was achieved by means of a two-layer ANN-RBF [60]. The network was trained with a spread parameter of 0.5 and a mean squared error goal of 0.01. SWR [126] was used for feature selection along with the leave-one-patient-out cross-validation method.

6.5.3   RESULTS AND DISCUSSION

Performance analysis of the proposed features when used individually is summarized in Table 6.3 for two situations: one with the C-ROIs centered at the actual centroids of the masses and the other with the C-ROIs centered at the randomly selected centroidal points. In the latter case, the results were averaged over the 10 obtained values. The performance achieved in terms of A_z indicates that each of the proposed features is able to discriminate between benign lesions and malignant tumors with A_z up to 0.7. For the classification stage, using the C-ROIs centered at the actual centroids and the estimated size of the masses, $A_z = 0.98$ (SE = 0.03) was achieved when the selected set of features was used. At least three out of the four proposed features were always automatically selected for each patient in the leave-one-patient-out procedure. This fact indicates that the proposed measures of angular correlation and trend complement one other. ROCKIT failed to converge for two of the 10 sets of discriminant values obtained using the C-ROIs centered at the randomly selected centroidal points; two additional random selections were performed and used for analysis. The obtained results are summarized in Table 6.3 in terms of the average and standard deviation of the A_z. The best result, $A_z = 0.69$ (SE = 0.04), was achieved when C_{AI} was used individually.

Table 6.3: Performance analysis of features for characterization of angular correlation and angular trend

Feature	Actual Centroid A_z (SE)	Randomly Selected Centroidal Points Average A_z ± Std
C_{AI}	0.69 (0.06)	0.69 ± 0.04
C_{AG}	0.70 (0.06)	0.68 ± 0.02
T_{AI}	0.62 (0.07)	0.65 ± 0.03
T_{AG}	0.69 (0.07)	0.68 ± 0.03

CAI, CAG: angular correlation measures; TAI, TAG: angular trend measures. The Az (and standard error, SE) values were estimated using ROCKIT [140]. The average Az and standard deviation (Std) were obtained using the 10 randomly selected centroidal points for each C-ROI.

The final A_z, achieved by performing the feature selection and the ANN-RBF classification steps in the leave-one-patient out procedure 10 times, was 0.99 ± 0.01. The number of selected features that were used as inputs to the ANN varied between three and four from one step to another. The good results obtained using the randomly selected centroidal points demonstrate the robustness of the features to inaccuracies in the extraction of the C-ROIs. Moreover, the results obtained in the present study are better than the results provided by other automatic methods reported in the literature [38, 58, 102, 129] and the previous related study described in Section 6.4, showing that contour-independent features, requiring only an estimate of the center and an approximate size of the mass, can lead to an effective method for the classification of breast masses in mammograms.

6.6   REMARKS

In this chapter, the design of new features for the characterization of texture patterns related to malignant tumors has been described. The possibility of extending statistical techniques to generate feature descriptors in a 2D circular domain representation of a mass was explored. Four novel features have been proposed to quantify correlation and trend in the radial (Study 1) and angular (Study 2) directions. The experimental results, obtained using 146 ROIs, including 126 benign lesions and 20 malignant tumors from the FFDM dataset, demonstrate that the proposed features can be used for classification of breast masses in mammograms without requiring the extraction of the precise contours of the masses. Further results on the application of the features described in this chapter in an integrated CADe/CADx system for mammography are presented in the next chapter.

CHAPTER 7

Integrated CADe/CADx of Mammographic Lesions

7.1 MOTIVATION

Detection and interpretation of signs of breast cancer in screening mammography are difficult tasks. Demands for reducing the number of unnecessary breast biopsies and obtaining higher sensitivity of detection of malignant tumors in mammographic screening indicate the need for new effective and comprehensive computerized solutions for mammography. In addition to assessing whether there is any breast abnormality and determining its location on the mammogram, a clinically effective CAD system should provide an estimate of the degree of malignancy of the detected lesion in order to prompt the radiologist for further assessment and treatment in the presence of pathological processes. This would facilitate accurate interpretation of mammograms and efficient treatment of breast cancer. The integration of detection and classification in a unified CAD system for mammography can increase the detection rate of breast cancer and reduce the FP recalls. Improved performance levels and robustness of the a CAD systems can be achieved if the detection and characterization of lesions, which underly the decision making processes, include the quantification of features related to malignancy of the lesions.

7.2 STATE OF THE ART

The state of the art provides a large number of techniques and approaches for the detection and characterization of lesions in mammograms, including calcifications, masses, bilateral asymmetry, and architectural distortion. A substantial number of successesful methods have been reported for automatic detection of breast calcifications, with levels of sensitivities up to 98.3% with 0.3 FP clusters per image [31]. Such systems are commercially available and used in clinical practice, with limited possibilities of improvement. However, this is not the case for masses, architectural distortion, and bilateral asymmetry. Studies on automatic identification of regions of architectural distortion reported sensitivies of detection in the range [0.68–0.93] at rates of [2.3–15] FPs per image [124], while the range of sensitivity of detection of masses reported for commercial CAD systems is [0.43–0.80] at a range of FPs of [0.2–1.4] per image [105]. A detailed analysis of the state of the art of CAD systems for detection of bilateral asymmetry and related innovative solutions has been presented in Chapter 5.

Automatic detection of masses and architectural distortion is challenging due to their subtle nature and similarity with the normal structures of the breast. CAD of breast cancer has been mostly addressed so far by dealing with automatic detection of benign and malignant masses, or architectural distortion, without distinction [105, 130, 147]. Although improvements to the detection tasks are still possible, classification of breast abnormalities in order to discriminate between malignant tumors and benign lesions is still an open issue. The characterization of automatically detected lesions in terms of malignancy, in particular, leads to further complexity as a result of the integration of methods for both detection and classification of the lesions in a unified system [32]. A few works in the literature only partly addressed this topic of research [38, 102, 116] and many challenges are still to be tackled for increased performance levels and transparency of the automated decision process [43].

7.3   OVERVIEW OF THE INTEGRATED CADE/CADX SYSTEM

The hypothesis that forms the basis of this work is that feature descriptors of lesions, which are needed for the detection of abnormalities in the breast parenchyma, need to be integrated with feature descriptors of malignancy which correlate, instead, to breast cancer. The approach proposed in this work includes the development of a clinically effective CAD system by combining two models of classification, one for the detection and one for the classification of the lesions. This implies the development of a novel comprehensive and multistage system for automatic detection and diagnosis of malignant tumors in a realistic scenario of a three-class space environment, i.e., in the presence of normal and abnormal mammograms including normal parenchymal tissue, benign lesions, and malignant tumors. The system has been designed so that accurate extraction of the contours of the lesions to be detected is not needed [23, 26]. This is an attempt to facilitate the detection of tumors that are conventionally considered to be difficult to identify by automated procedures.

The following main contributions and their integration in a unified CAD system for mammography will be presented with more details in the sections of this chapter.

1. Analysis of the Gaussian and mean curvatures of the mammographic appearance of the breast region, which is represented as a three-dimensional (3D) surface of intensity values as a function of spatial coordinates, i.e., a Monge patch, and extraction of suspicious focal areas based on cardinality restrictions on the average Gaussian curvature values.

2. Rejection of the oriented structures of the normal breast parenchyma, such as portions of skin-line, vessels, and breast ducts, via analysis of the phase response of multidirectional Gabor filters.

3. Extraction of C-ROIs and subsequent adaptive determination of annular regions including lesion candidates.

4. Use of a differential approach to the reduction of falsely detected lesions that is based on the quantification of the differences in texture and density between the region of a detected candidate and the surrounding tissue.

5. Use of the contour-independent features described in Chapter 6 for the classification of automatically detected malignant tumors, by including space and intensity as explanatory variables to analyze the local nonstationarity [7] and spatial dependence [83] of pixel values in the angular and radial direction starting from circular or annular regions depicting a lesion.

6. Introduction of a unified 3D FROC characteristic framework for evaluation and analysis of two binary classification problems in series, the detection of lesions and the classification of malignant tumors, with which to assess the cumulative performance and error of a CAD system in detecting breast cancer with respect to the relative number of falsely detected tumors, including benign lesions and normal tissue.

7. Validation of the proposed CAD system on a diversified set of images including FFDMs and SFMs, extracted from three different databases and acquired with different devices: a private set of images collected at the San Paolo Hospital of Bari and two publicly available databases, the DDSM [61] and the mini-MIAS [141] database.

A flowchart of the algorithm is illustrated in Fig. 7.1 for an overview of the main methodologies used in this study.

7.4   DATASETS AND EXPERIMENTAL SETUP

Two expert radiologists manually marked the contours of 146 masses in 138 FFDMs of 88 patients. All of the available FFDMs with marked contours of lesions were included in this study. The contours were used, at first, to extract ROIs including the masses in a preliminary study design of contour-independent features for classification of lesions (see Chapter 6). Results of biopsy indicated 120 masses as being benign and 26 as malignant tumors. This information was used as the ground-truth for the diagnostic classification. The same contours were used as ground truth for the detection task. All of the available normal mammograms, 18 images in total, were also included for analysis, thus obtaining a dataset of 76 CC and 80 MLO views for validation of the integrated CADe/CADx system.

Publicly available SFMs were also included in the analysis to facilitate comparison of the results obtained in this work with other reported methods for the analysis of mammograms. With this purpose, all DDSM cases with at least one mass or one region of architectural distortion digitized with the Lumisys 200 laser densitometer were included in the study, for a total of 1688 images (422 patients). The images have a spatial resolution of 50 and grayscale resolution of 12 bpp. A summary of the dataset is provided in Table 7.1, where the number of lesions (and malignant tumors) is indicated in relation to the margins and the BI-RADS density together

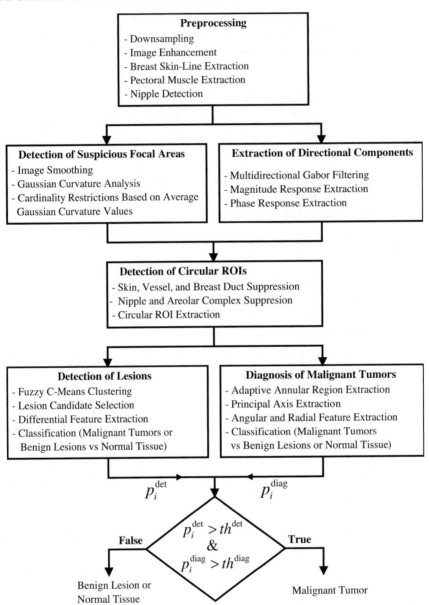

Figure 7.1: Flowchart of the proposed CAD system for detection and classification of lesions in mammograms. The probabilities of detection, p_i^{det}, and diagnosis, p_i^{diag}, represent the probability of being a lesion and of being a malignant tumor, respectively. The two probabilities are combined together using thresholds for detection, th^{det}, and diagnosis, th^{diag}. Reproduced with permission from Casti et al. [26] ©Elsevier.

Table 7.1: Numbers of lesions (malignant tumors) and normal mammograms in terms of margins and BI-RADS density (B-I to B-IV) [39] for the DDSM dataset [61]

	B-I	B-II	B-III	B-IV	Total
Masses					
Circumscribed	82 (18)	150 (14)	64 (3)	9 (0)	305 (35)
Microlobulate	36 (34)	42 (24)	17 (11)	0	95 (69)
Spiculated	43 (41)	81 (81)	27 (26)	3 (3)	154 (151)
Obscured/Ill-defined	48 (30)	124 (67)	109 (48)	45 (14)	316 (159)
Other	1 (1)	3 (1)	3 (0)	0	7 (2)
Total	210 (124)	400 (187)	220 (88)	57 (17)	877 (416)
	B-I	**B-II**	**B-III**	**B-IV**	**Total**
Architectural Distortion					
Microlobulate	0	2 (2)	0	0	2 (2)
Spiculated	6 (5)	28 (22)	15 (15)	4 (2)	54 (44)
Obscured/Ill-defined	4 (4)	12 (9)	29 (25)	7 (7)	52 (45)
Other	1 (1)	5 (3)	7 (3)	2 (0)	15 (7)
Total	11 (10)	47 (36)	51 (43)	13 (9)	122 (98)
	B-I	**B-II**	**B-III**	**B-IV**	**Total**
Normal Mammograms	152	358	231	44	785

Table 7.2: Numbers of masses and regions of architectural distortion in terms of biopsy result and BI-RADS density [39] for the DDSM dataset [61]

	Malignant		Benign		Benign Without Callback		Unproven	
	Masses	Arch. Dist.	Masses	Arch. Dist.	Masses	Arch. Dist.	Masses	Arch. Dist.
B-I	124	10	78	0	4	1	4	0
B-II	187	36	207	6	2	4	4	1
B-III	88	43	124	8	0	0	8	0
B-IV	17	9	28	4	0	0	2	0
Total	416	98	437	18	6	5	18	1

with the density of the normal mammograms. In Table 7.2, the number of lesions is reported in terms of margins, biopsy results, and BI-RADS density [39]. An example of a mammogram including a benign mass and a region with malignant architectural distortion is illustrated in Fig. 7.2b.

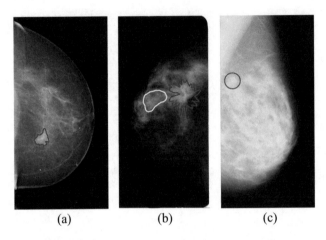

(a)   (b)   (c)

Figure 7.2: (a–c) Examples of mammograms including a lesion. The red and yellow contours indicate the regions annotated by radiologists. (a) Case $CD14\_PA8\_ST1\_LEFT\_MLO$ from the FFDM dataset including a malignant tumor. (b) Case $B\_3036\_1.RIGHT\_CC$ from the DDSM including a benign mass, in yellow, and a malignant architectural distortion, in red. (c) Case $mdb030$ from the MIAS dataset including a malignant tumor. Reproduced with permission from Casti et al. [26] ©Elsevier.

In addition, a third set of mammograms was extracted from the MIAS [141] database. In order to facilitate comparison of the results obtained in this study with the other results reported in the literature, all cases from the MIAS database analyzed in the review by Oliver et al. [105] were selected, where seven previously published approaches were selected as being representative of the main state-of-the-art techniques for detection of masses [44, 70, 77, 80, 114, 116, 153]. The dataset consists of 261 MLO mammograms, of which 207 are normal and 54 contain masses (see the example in Fig. 7.2c). A summary of the dataset is provided in Table 7.3 in terms of margins and BI-RADS density, and in relation to the biopsy results and the BI-RADS density in Table 7.4. The normal mammograms are also summarized in Table 7.3. The annotated lesions and corresponding biopsy results were used for validation of the various stages of the system.

Table 7.3: Numbers of lesions (malignant tumors) and normal mammograms in terms of margins and BI-RADS density (B-I to B-IV) [39] for the MIAS dataset [141]

	B-I	B-II	B-III	B-IV	Total
Circumscribed	9 (2)	6 (1)	3 (1)	2 (0)	20 (4)
Spiculated	4 (1)	7 (5)	8 (1)	1 (1)	20 (8)
Obscured/Ill-defined	7 (5)	4 (3)	3 (0)	0	14 (8)
Total	20 (8)	17 (9)	14 (2)	3 (1)	54 (20)
Normal Mammograms	56	67	58	26	207

Table 7.4: Numbers of masses and regions of architectural distortion in terms of biopsy result and BI-RADS density [39] for the MIAS dataset [141]

	Malignant Masses	Benign Masses
B-I	8	12
B-II	9	8
B-III	2	12
B-IV	1	2
Total	20	34

7.5   METHODS

7.5.1   PREPROCESSING

For computational reasons and uniformity, all the images were downsampled to the spatial resolution of 200 using a bicubic interpolation method. Due to the variability of the images analyzed, differentiated contrast enhancement procedures were applied to each dataset: a linear transformation as encoded in the GE look-up table (LUT) for the FFDMs and a power-law transformation for the DDSM dataset. Mammograms from the MIAS dataset were left unaltered. The main landmarking structures, i.e., the breast skin-line, the nipple, and the pectoral muscle (only for MLO views), were extracted by the methods described in Chapter 4 and used for extraction of the breast region, as shown in Figs. 7.3a, 7.4a, and 7.5a for the examples in Figs. 7.2a, b, and c. Although the choice of removing the pectoral muscle (for MLO views) from the breast region to be analyzed may cause the loss of some lesions, it was necessary due to the presence of intricate anatomic structures or brighter areas that may affect the detection performance by causing FP outcomes. Dedicated procedures would be required for optimal analysis and evaluation of the pectoral muscle regions in MLO views.

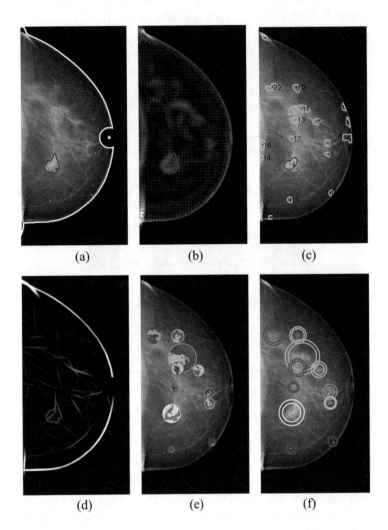

Figure 7.3: Main steps of the algorithm for the detection and diagnosis of malignant tumors for an example from the FFDM. The red contour indicates the region annotated by the radiologist. (a) Original mammogram and landmarking structures: breast skin-line, pectoral muscle, nipple-areolar complex. (b) GVF. (c) The 20 suspicious focal areas automatically detected via μ_K. The marked points indicate the centers of the lesion candidates used for evaluation of results and ranked in descending order of μ_K. (d) Gabor magnitude response. (e) C-ROIs and corresponding clustering. (f) Circular and annular regions used for the classification task. Reproduced with permission from Casti et al. [26] ©Elsevier.

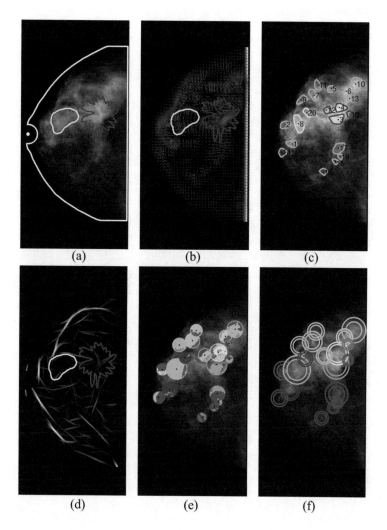

Figure 7.4: Main steps of the algorithm for the detection and classification of lesions for an example from the DDSM. The red and yellow contours indicate the regions annotated by the radiologist. (a) Original mammogram and landmarking structures: breast skin-line, pectoral muscle, nipple-areolar complex. (b) GVF. (c) The 20 suspicious focal areas automatically detected via μ_K. The marked points indicate the centers of the lesion candidates used for evaluation of results and ranked in descending order of μ_K. (d) Gabor magnitude response. (e) C-ROIs and corresponding clustering. (f) Circular and annular regions used for the classification task. Reproduced with permission from Casti et al. [26] ©Elsevier.

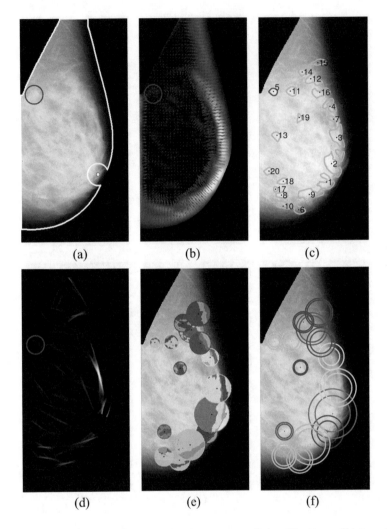

Figure 7.5: Main steps of the algorithm for the detection and classification of lesions for an example from the MIAS dataset. The red contour indicates the region annotated by the radiologist. (a) Original mammogram and landmarking structures: breast skin-line, pectoral muscle, nipple-areolar complex. (b) GVF. (c) The 20 suspicious focal areas automatically detected via μ_K. The marked points indicate the centers of the lesion candidates used for evaluation of results and ranked in descending order of μ_K. (d) Gabor magnitude response. (e) C-ROIs and corresponding clustering. (f) Circular and annular regions used for the classification task. Reproduced with permission from Casti et al. [26] ©Elsevier.

7.5.2   DETECTION OF SUSPICIOUS FOCAL AREAS

Gaussian filtering of the images with $\sigma = 12$ pixels (2.4 mm) was performed to reduce noise and retain only low-frequency information related to the cores of the lesions. The chosen value of σ preserves lesions of size 2.3–2.6 mm, which is reported as the minimum size required to detect a lesion on mammograms [138].

The intensity values of a given image can be characterized in terms of topology in $\mathbb{R}^3$ via the definition of a Monge patch, $\xi : U \longmapsto \mathbb{R}^3$, as

$$\xi(u, v) = [u, v, h(u, v)], \tag{7.1}$$

where U is the domain set in $\mathbb{R}^2$ containing the two space variables, u and v, and $h : U \longmapsto [0, 1]$ is a function that associates with each pair of spatial coordinates (u, v) an intensity value $h(u, v)$ [56]. The Gaussian curvature, K, defined as the product of the two principal curvatures of a surface at any of its points determines the topology of any given surface. Let $\mathbf{p} = (u_0, v_0)$ be the coordinates of a generic point on ξ; then, the topology classification can be performed as follows.

- If $K(u_0, v_0) > 0$, $\mathbf{p}$ is elliptic.

- If $K(u_0, v_0) < 0$, $\mathbf{p}$ is hyperbolic.

- If $K(u_0, v_0) = 0$, $\mathbf{p}$ is parabolic or planar.

The conditions stated above can be reformulated by considering that, for the class of Monge patches, K assumes the form

$$K = \frac{h_{uu} h_{vv} - h_{uv}^2}{\left(1 + h_u^2 + h_v^2\right)^2}, \tag{7.2}$$

embedding the first (h_u and h_v) and second (h_{uu}, h_{vv}, and h_{uv}) partial derivatives of h. Recalling the codomain of the first derivatives of h, $[-1, 1]$, and the smoothing of the high-frequency details in the image resulting from the initial step of Gaussian filtering, $h_u^2 + h_v^2 \ll 1$ is considered; then,

$$K \simeq \det \left[\text{Hess}(h)\right]. \tag{7.3}$$

If an orthonormal basis of eigenvectors for $\text{Hess}(h)$ is chosen, K can also be expressed via the product of the corresponding eigenvalues, λ_1 and λ_2. It is hypothesized that suspicious areas on a mammogram satisfy the constraint

$$\forall\, (u, v) \in U : \{\lambda_{1,2}(u, v) < 0\}, \tag{7.4}$$

where the concordance of the signs of λ_1 and λ_2 results in positivity of K, while the negative values for λ_1 and λ_2 correspond to negative values of h_{uu} and h_{vv}, resulting in negative concavities of both the u and v cross sections. Such a condition leads to the selection of a set of focal areas on the mammogram that correspond to the nodes of a convergent GVF. The average value of K, labeled as μ_K, of pixels within each selected region was computed and used for rank-ordering the focal areas up to a maximum of 20 candidates per image, as shown in Figs. 7.3c, 7.4c, and 7.5c. For each region, the point c_f with the highest value of K was taken as the corresponding center.

Since the initial average area of the regions on a mammogram satisfying the conditions on the eigenvalues was found to be about 3000 pixels (120 mm^2), selecting 20 candidates per image corresponds to evenly covering a square region of the same area with 20 ROIs of side about 11 mm, which is the average size of an invasive carcinoma detected by mammography [78].

7.5.3   EXTRACTION OF DIRECTIONAL COMPONENTS

A set of 18 multidirectional Gabor filters equally spaced over the angular range $(-\pi/2, \pi/2]$ was used to extract the various directional components of the breast region to be analyzed, as described in Section 3.1. The elongation parameter, l, was set to 3 while a thickness of 24 pixels was set as the design parameter, τ. The obtained Gabor magnitude response is illustrated in Figs. 7.3d, 7.4d, and 7.5d.

7.5.4   EXTRACTION OF CIRCULAR ROIS

Suspicious focal areas having pixels with fewer than six different orientations were rejected. This procedure helped in rejecting erroneously selected breast ducts, vessels, and portions of the skin-line, such as regions 16 and 19 in Fig. 7.3c. The nipple-areolar complex was masked by means of a disk of radius 50 pixels (10 mm) centered at the detected position of the nipple [8] and all of the detected regions whose center belonged to this region were rejected, as for the case of candidate 8 in Fig. 7.3c. The radius of each C-ROI was determined as the maximum radial distance of the center of each detected suspicious focal area, c_f, from the corresponding boundary, as shown in Figs. 7.3e, 7.4e, and 7.5e.

7.5.5   EXTRACTION OF FEATURES FOR DETECTION OF LESIONS

The detected C-ROIs were first analyzed in order to discriminate lesions from normal tissue. For this purpose, the suspicious focal areas, a different representation of the lesion candidates, was created in order to include additional parenchymal tissue around each candidate. Differences in the geometry and oriented patterns between the two representations were quantified by means of differential features, which are expected to capture information on the presence of tissue abnormalities.

The pixels within each C-ROIs were grouped into four different clusters via the fuzzy c-means data clustering technique [9], in relation to the following structures [1]: lobules or glandular tissue, ducts, fatty tissue, and fibrous connective tissue. Examples of the obtained clusters are presented in Figs. 7.3e, 7.4e, and 7.5e; the color code indicates the obtained clusters. For each C-ROI, the cluster including the pixel of interest c_f was selected to be analyzed.

    The selected regions, together with the original suspicious focal areas previously detected, were used to extract a set of features for quantification of the difference between the two representations: the detected focal area, indicated by the subscript f, and the corresponding cluster, indicated by the subscript c. Seven differential features, i.e., area (A and ΔA), compactness (k and Δk), elongation (E and ΔE), distance (d), directionality (φ), weighted directionality (φ_w), and dispersion (s), were computed as follows:

$$\Delta A = |A_c - A_f| / (A_c + A_f), \tag{7.5}$$

$$\Delta k = |k_c - k_f|, \tag{7.6}$$

$$\Delta E = |E_c - E_f|, \tag{7.7}$$

$$d = \|c_c - c_f\|, \tag{7.8}$$

$$\varphi = |std[\Phi_c(u, v)] - std[\Phi_f(u, v)]|, \tag{7.9}$$

$$\varphi_w = |std[(\Phi_c \circ M_c)_{u,v}] - std[(\Phi_f \circ M_f)_{u,v}]|, \tag{7.10}$$

$$s = std[\|B_c(u, v) - c_f\|] / mean[\|B_c(u, v) - c_f\|], \tag{7.11}$$

where E is defined as the ratio between the major and the minor axes of the ellipse that has the same normalized second central moments as the region; Φ and M correspond to the phase and magnitude responses of the multidirectional Gabor filters, respectively; d is the Euclidean distance between the centers of the detected cluster, c_c and the focal area, c_f; std indicates the standard deviation; $\circ$ denote the pixel-by-pixel multiplication; and $B_c(u, v)$ are the boundary pixels of the cluster. The correspondence of the two representations, and hence lower values of the differential features, should indicate the presence of a focal lesion that retains a certain degree of consistency with respect to the surrounding structures. On the contrary, higher values of the differential features should characterize lobules and fibroglandular tissue which, instead, are expected to extend beyond the focal zones initially detected showing, with respect to them, variations in geometrical and directional structures. In addition, for each candidate, the value of μ_K, and the normalized values for the maximum and standard deviation of the values of μ_K of all of the candidates detected on the mammogram being analyzed, were computed.

7.5.6   EXTRACTION OF FEATURES FOR CLASSIFICATION OF LESIONS

Diagnostic classification of the detected C-ROIs as malignant tumors against benign lesions or normal tissue was performed by means of the contour-independent features described in Chapter 6. The features were computed from the detected C-ROIs and from the ribbons surrounding them (see the regions in Figs. 7.3f, 7.4f, and 7.5f). The width of the circular ribbons was adaptively determined as $R_w = R(A_f/A_r)$ [102], where R and A_r are the radius and area of the C-ROI, respectively, and A_f is the area of the corresponding focal region previously detected. The proposed features were computed from the gray-scale values and the Gabor magnitude response values to derive information on the spatial distribution of the density and oriented patterns, respectively.

7.5.7 PATTERN CLASSIFICATION AND CROSS-VALIDATION

The three-class classification problem with the classes normal tissue, benign lesion, and malignant tumor was decomposed into a series of two two-class classification stages, each consisting of SWR for selection of features and an FLDA classifier. The first stage of classification addressed the discrimination of lesions from normal tissue by using the differential features (see Section 7.5.5), whereas the discrimination of malignant tumors from benign lesions or normal tissue was performed by the second stage using the contour-independent features (see Section 7.5.6). The cases from each dataset were randomly split into two subgroups of equal size for training and testing the two classifiers (two-fold cross-validation). SWR was performed only on the training sets. The procedure was repeated twice for each dataset by exchanging the two subgroups in order to derive, in the test phases, the probabilities of detection, p_i^{det}, and diagnosis, p_i^{diag}, for all of the C-ROIs.

7.5.8 PERFORMANCE EVALUATION

Evaluation of the performance achieved by the automatically selected features at each of the two stages of classification was performed using the area under the receiver operating characteristic (ROC) curve (A_z). A lesion candidate was considered as a true positive (TP) in detection and a TP in classification if the corresponding center, c_f, was, respectively, within an annotated lesion and within a malignant tumor. Additional TPs within the same annotated region were not counted. FP candidates in detection corresponded to normal tissue regions erroneously detected, whereas FP candidates in classification accounted for both normal tissue regions and benign lesions incorrectly classified as being malignant.

7.5.9 3D FROC FRAMEWORK

FROC analysis was performed by means of 101 discriminant values equally spaced over the range [0, 1], which were used as thresholds for detection, th^{det}, and diagnosis, th^{diag}. A unified 3D FROC framework was used to represent the cumulative performance and error of the workflow for the detection and classification of breast cancer with respect to the falsely detected tumors [24]. The indices of performance, i.e., TPR and FPpI, assume different meaning depending on the stage of classification. For the detection of lesions only, TPR corresponds to the ratio of the number of detected lesions to the total number of lesions, including masses and regions of architectural distortion, whereas FPpI refers to the number of falsely detected lesions divided by the total number of images. For the classification stage, TPR, also referred to as sensitivity of detection of malignant tumors, corresponds to the number of malignant tumors detected and classified as being malignant divided by the total number of malignant tumors actually present. The corresponding FPpI, also named number of falsely detected malignant tumors per image, indicates the number of normal tissue regions and benign lesions incorrectly detected and classified as being malignant divided by the total number of images. The values of p_i^{det} and p_i^{diag} estimated for each C-ROI at the two independent stages of classification were combined together, at various combinations of thresholds using the AND logic operation, as shown in Fig. 7.1, to determine the final assessment. By varying the pairs of threshold values, $(th^{\mathrm{det}}, th^{\mathrm{diag}})$, the 3D FROC framework defines the values of sensitivity of detection of malignant tumors of the whole system and the corresponding number of falsely detected malignant tumors.

7.6 RESULTS AND COMPARATIVE ANALYSIS

7.6.1 INITIAL PERFORMANCE ASSESSMENT

The initial FROC curves for the detection of lesions for which the sensitivity of detection of lesions varies with the chosen maximum number of initial candidates per image are illustrated in Fig. 7.6. TPRs (and number of detected lesions) of 0.96 (140), 0.93 (908), and 0.94 (51) were obtained for the FFDM, DDSM, and MIAS datasets, respectively, at FPpI values of 18.84,

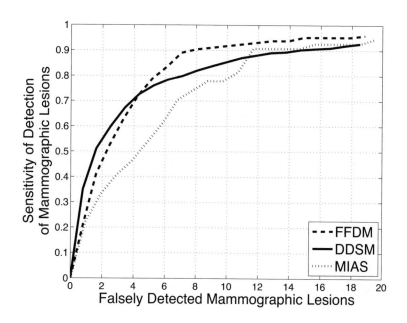

Figure 7.6: FROC curves of the initial detection of suspicious focal areas for the three datasets of mammograms. Reproduced with permission from Casti et al. [26] ©Elsevier.

18.51, and 19.49 by the initial candidate selection procedure; the corresponding total number of initial candidates was 3079, 32147, and 5138, respectively. Among the lesions counted as missed by using the previously mentioned criterion of the detected center, 2 masses for the FFDM, 23 lesions for the DDSM, and 2 masses for the MIAS had more than 50% of the corresponding areas delineated by the radiologists outside the analyzed breast regions, mostly hidden by the extracted pectoral muscle. The number of C-ROIs used for training the classifiers was 2057, 25816, and 3765, for the FFDM, DDSM, and MIAS datasets, respectively, to which corresponded initial sensitivities of detection (and number of detected lesions) of 0.93 (136), 0.89 (889), and 0.93 (50) at 12.10, 14.32, and 14.13 falsely detected lesions per image. The sensitivities of detection of malignant tumors (and number of detected malignant tumors) were, in order, 0.92 (24), 0.94 (481), and 0.95 (19) at 12.95, 14.65, at 14.31 falsely detected malignant tumors per image.

7.6.2 PERFORMANCE OF CLASSIFICATION

The two sets of features selected by SWR achieved, in the test phases of the two FLDA classifiers, $A_z = 0.83$ and $A_z = 0.80$ for the detection and diagnosis of masses and malignant tumors, respectively, for the 156 mammograms of the FFDM dataset. For the 1688 mammograms of the DDSM dataset, ROC analysis provided an A_z value of 0.72 in distinguishing masses and regions of architectural distortion from the normal candidates; the diagnosis of malignant tumors from the rest of the candidates (benign lesions and normal tissue) was achieved with $A_z = 0.61$. For the case of the 261 mammograms from the MIAS database, A_z values of 0.73 and 0.85 were obtained, respectively, for the detection of masses and diagnosis of malignant tumors.

The influence of the margins of the lesions and the BI-RADS density on the performance of the individual stages of classification was investigated. The obtained results are presented in Table 7.5 in terms of A_z. The algorithm shows good performance in the detection and diagnosis of obscured or ill-defined tumors. Lesions in fatty breasts, B-I, led to better results; all the classes of density showed satisfactory to good performance. However, this preliminary ROC analysis of the performance of each of the two classification stages individually does not take into account the combination of detection and diagnosis performed by the CAD system for the final assessment of the lesions.

The analysis of the 3D FROC framework provided the values of TPR for detection and diagnosis at the corresponding levels of FPpI. Figures 7.7a, 7.8a, and 7.9a show the FROC curves obtained for the diagnosis of malignant tumors for the FFDM, DDSM, and MIAS datasets, respectively. The diagnosis threshold values, indicated by the color code, are combined at the various detection threshold values, indicated along the y-axis, to obtain pairs of TPR and FPpI. The falsely detected malignant tumors indicated along the x-axis account for both normal tissue regions and benign lesions when incorrectly classified as being malignant. As expected, lower values of both the detection and the diagnosis thresholds lead to higher TPR at

Table 7.5: Values of A_z for detection of lesions and diagnosis of malignant tumors in terms of margins and BI-RADS density (B-I to B-IV) [39]. Results were obtained individually by each of the two stages of classification consisting of SWR for selection of features, FLDA, and two-fold cross-validation. Values higher than 0.8 are in bold.

	Detection		Diagnosis	
	DDSM	MIAS	DDSM	MIAS
Circumscribed	0.72	0.68	0.64	0.67
Microlobulated	**0.81**	–	0.61	–
Spiculated	0.73	0.73	0.61	**0.84**
Obscured/Ill-defined	0.72	**0.81**	0.59	**0.92**

(a) Margins

	Detection		Diagnosis	
	DDSM	MIAS	DDSM	MIAS
B-I	0.78	**0.89**	0.63	**0.88**
B-II	0.72	0.71	0.61	**0.83**
B-III	0.69	0.71	0.57	0.63
B-IV	0.66	0.65	0.63	–

(b) BI-RADS Density

the expense of increased FPpI. On the contrary, values of the detection and diagnosis thresholds close to unity reduce the TPR to zero.

The projection planes along the y direction of the 3D FROC plots are shown in Figs. 7.7b, 7.8b, and 7.9b for the FFDM, DDSM, and MIAS datasets, respectively, in order to illustrate the operating regions obtained in the standard FROC plane. At each level of sensitivity of detection of malignant tumors, it is possible to determine the best detection and diagnosis thresholds so that the number of falsely detected malignant tumors is minimized. The values of FPpI obtained for the three datasets of mammograms are summarized in Tables 7.6, 7.7, and 7.8 for both the detection and the diagnosis phases at FPpI of about 0.70, 0.80, and 0.90, together with the number of correctly detected and falsely detected lesions. Overall, the system detected 80% of malignant tumors from the DDSM and MIAS datasets at 3.47 and 2.92 falsely detected malignant tumors per image, respectively. For the FFDM dataset, a sensitivity of 0.81 corresponded to 1.04 falsely detected malignant tumors per image.

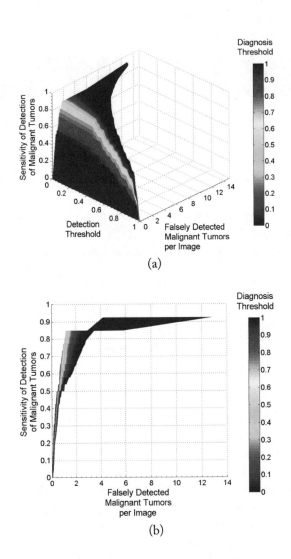

Figure 7.7: 3D-FROC analysis framework for global optimization of detection and diagnosis of malignant tumors for the FFDM dataset. The detection and diagnosis thresholds are indicated along the x-axis and by the color code map, respectively. (a) 3D FROC. (b) 3D FROC projected onto the y-z plane. Reproduced with permission from Casti et al. [26] ©Elsevier.

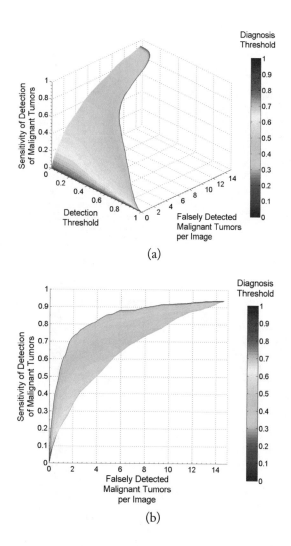

(a)

(b)

Figure 7.8: 3D-FROC analysis framework for global optimization of detection and diagnosis of malignant tumors for the DDSM dataset. The detection and diagnosis thresholds are indicated along the x-axis and by the color code map, respectively. (a) 3D FROC. (b) 3D FROC projected onto the y-z plane. Reproduced with permission from Casti et al. [26] ©Elsevier.

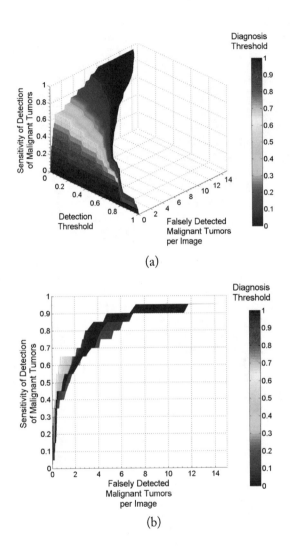

(a)

(b)

Figure 7.9: 3D-FROC analysis framework for global optimization of detection and diagnosis of malignant tumors for the MIAS dataset. The detection and diagnosis thresholds are indicated along the x-axis and by the color code map, respectively. (a) 3D FROC. (b) 3D FROC projected onto the y-z plane. Reproduced with permission from Casti et al. [26] ©Elsevier.

Table 7.6: Results of 3D FROC analysis for detection of lesions and diagnosis of malignant tumors obtained by combining two sets of features selected by SWR, a series of two FLDA classifiers and two-fold cross-validation for the FFDM dataset

		Detection				Diagnosis		
CC/MLO	TPR (TP/# Lesions)	0.71 (104/146)	0.80 (117/146)	0.90 (132/146)	TPR (TP/# Tumors)	0.73 (19/26)	0.81 (21/26)	0.92 (24/26)
	FP/Image (FP/# Images)	2.19 (341/156)	2.94 (459/156)	7.00 (1093/156)	FP/Image (FP/# Images)	0.82 (128/156)	1.04 (162/156)	4.02 (627/156)
CC	TPR (TP/# Lesions)	0.70 (49/70)	0.80 (56/70)	0.90 (63/70)	TPR (TP/# Tumors)	0.75 (9/12)	0.83 (10/12)	0.92 (11/12)
	FP/Image (FP/# Images)	1.99 (151/76)	2.42 (184/76)	3.96 (301/76)	FP/Image (FP/# Images)	0.57 (43/76)	0.70 (53/76)	0.86 (65/76)
MLO	TPR (TP/# Lesions)	0.71 (54/76)	0.80 (62/76)	0.87 (69/76)	TPR (TP/# Tumors)	0.71 (10/14)	0.79 (11/14)	0.86 (12/14)
	FP/Image (FP/# Images)	2.39 (191/80)	3.44 (275/80)	12.05 (964/80)	FP/Image (FP/# Images)	0.88 (70/80)	1.19 (95/80)	3.76 (301/80)

Table 7.7: Results of 3D FROC analysis for detection of lesions and diagnosis of malignant tumors obtained by combining two sets of features selected by SWR, a series of two FLDA classifiers and two-fold cross-validation for the DDSM dataset

		Detection				Diagnosis		
CC/MLO	TPR (TP/# Lesions)	0.70 (700/999)	0.80 (800/999)	0.88 (880/999)	TPR (TP/# Tumors)	0.73 (360/514)	0.80 (412/514)	0.90 (463/514)
	FP/Image (FP/# Images)	3.42 (5776/1688)	6.20 (10461/1688)	12.06 (20358/1688)	FP/Image (FP/# Images)	1.68 (2831/1688)	3.47 (5856/1688)	8.25 (13933/1688)
CC	TPR (TP/# Lesions)	0.70 (350/499)	0.80 (400/499)	0.89 (435/499)	TPR (TP/# Tumors)	0.70 (181/257)	0.81 (208/257)	0.90 (232/257)
	FP/Image (FP/# Images)	2.56 (2163/844)	4.87 (4109/844)	10.40 (8775/844)	FP/Image (FP/# Images)	1.54 (1299/844)	2.43 (2049/844)	5.77 (4871/844)
MLO	TPR (TP/# Lesions)	0.70 (351/500)	0.80 (400/500)	0.87 (445/500)	TPR (TP/# Tumors)	0.70 (180/257)	0.80 (206/257)	0.90 (232/257)
	FP/Image (FP/# Images)	4.04 (3408/844)	7.84 (6617/844)	13.13 (11078/844)	FP/Image (FP/# Images)	2.18 (1836/844)	4.18 (3527/844)	10.88 (9182/844)

Table 7.8: Results of 3D FROC analysis for detection of lesions and diagnosis of malignant tumors obtained by combining two sets of features selected by SWR, a series of two FLDA classifiers and two-fold cross-validation for the MIAS dataset

	Detection			Diagnosis			
MLO	TPR (TP/# Masses)	0.70 (38/54)	0.81 (44/54)	0.91 (49/54)	TPR (TP/# Tumors)	0.70 (14/20)	0.80 (16/20)	0.90 (18/20)
	FP/Image (FP/# Images)	2.12 (563/261)	4.02 (1051/261)	8.75 (2284/261)	FP/Image (FP/# Images)	2.16 (563/261)	2.92 (761/261)	4.52 (1179/261)

7.6.3   COMPARATIVE ANALYSIS

Due to the use of different datasets of mammograms and variations in the methods for evaluation of results, objective comparison of the performance of the present CAD system with the state of the art of CAD systems for mammography is difficult to perform. However, it is possible to report a comparison of the results achieved in this study with the results obtained by Oliver et al. [105] on the MIAS dataset by applying seven previously developed methods [44, 70, 77, 80, 114, 116, 153]. Among them, a sensitivity of 0.80 was reached only by four approaches [70, 80, 114, 153] at FPpI in the ranges of [3.96–5.42], [3.11–5.17], [5.97–6.58], and [5.92–6.61]. In this work, on the same set of mammograms, 4.02 FPpI at a sensitivity of detection of masses of 0.81 was obtained. When detection and diagnosis of malignant tumors were combined together, 2.92 FPpI at a sensitivity of detection of malignant tumors of 0.80 was achieved, which is better than the results reported for the other approaches, especially with reference to the ill-defined lesions (see Table 7.5).

The results reported for commercial CAD systems range between 0.43 and 0.80 for the sensitivity of detection of masses for the range of FPpI of [0.2–1.4] [105]. If the results obtained in this study with the FFDM dataset are considered, our methods outperform the commercial CAD systems evaluated with the additional feature of diagnosis of malignant tumors. The values of sensitivity of detection of malignant tumors obtained with the two SFM datasets are higher than the values reported for commercial CAD systems, at the expense of increased values of falsely detected malignant tumors per image.

7.7   DISCUSSION

Automatic detection and diagnosis of malignant tumors in mammograms is still an open issue due to their indistinct and diverse nature. Accurate extraction of indistinct and ill-defined boundaries of obscured lesions is difficult to achieve. The problem is made more difficult by the superposition of the surrounding parenchyma on the margins of masses in radiographically dense breasts. In this study, the development of a novel contour-independent CAD scheme for mammographic lesions was addressed. The analysis of the curvatures and of the directional components, together with the use of differential features for detection and contour-independent angular and radial features for diagnosis overcome the limitations of previous approaches due to reliance on accurate extraction of the contours of the lesions. Such limitations, deriving from the intrinsic nature of contours of lesions as they appear on the mammogram, which is inherently fuzzy, emerge directly from the intraexpert and interexpert variability in depicting contours of lesions. The proposed system integrates two main complex tasks: the identification of lesions on mammograms and the classification of the detected lesions as malignant tumors or benign lesions. The operating point, i.e., the pair (TPR, FPpI), can be set by the user in relation to the prevalence of the disease or the specific task to be addressed via analysis of the 3D FROC framework; this automatically determines the optimal setting of the cut-off values for both the detection and the diagnosis tasks. The results obtained through a comprehensive approach to the analysis of lesions indicate that detection and diagnosis can be combined together to increase the performance levels of CAD systems for mammography, as shown in Tables 7.6, 7.7, and 7.8.

With respect to the previous work on detection of masses by Mencattini and Salmeri [93], the number of empirical parameters of the related algorithm has been substantially reduced and the use of external inputs for setting the system, such as the BI-RADS mammographic density values, has been removed. Although these aspects may have caused a decrease in the obtained performance level, such conditions were required to increase both the generalizability of the approach to the various types of mammograms, FFDMs and SFMs, and the reliability of the reported results.

The proposed system is effective with FFDM, which plays a major role in the future of breast imaging. In addition, comparison of the results obtained in this study for the case of the SFMs of the MIAS dataset with the results obtained by other reported methods, when applied on the same set of mammograms, showed the efficacy and peculiarity of our system. The proposed comprehensive approach and the related global performance evaluation provide, for the first time, a unified CAD system for fully automated detection and diagnosis of malignant tumors in mammograms. Additional strategies can still be incorporated to reduce the FPRs and reach increased performance levels. In this study, for example, the simultaneous presence of microcalcifications as an indicator of malignancy was not investigated; neither was the integration of bilateral, ipsilateral, or temporal comparison [25]. Such additional features and combination of the proposed methods with other existing approaches for the analysis of mammograms are expected to bring improved sensitivity levels and to further reduce the FPR [105].

7.8   REMARKS

In this chapter, a novel comprehensive scheme to discriminate malignant tumors from benign lesions and normal parenchymal tissue has been presented in a realistic scenario of lesion candidates automatically detected on mammograms. The motivating factors for the development of an integrated CADe/CADx system for mammography have been described, together with a review of the related relevant literature. The results obtained on a total of 2105 FFDMs and SFMs from three different databases, including masses and regions of architectural distortion, demonstrate that detection and diagnosis can be combined together to increase the sensitivity of detection of malignant tumors while maintaining low FPRs.

Concluding Remarks

Mammographic screening programs followed by appropriate treatment is the established strategy to reduce breast cancer mortality. Due to the low prevalence of the disease in the screening population and due to difficulties in mammographic interpretation, CAD systems are currently used to reduce oversight errors. However, the performance levels of such systems and the related algorithmic solutions are not exhaustive.

In this book, new methods and innovative approaches for computerized diagnosis of breast cancer via mammography have been presented. The preliminary and important task addressed was the development of landmarking algorithms for detection of the anatomical components of the breast on mammograms, including the breast skin-line, the nipple, and the pectoral muscle. This is a preliminary step needed for most computerized systems for mammography. For the case of bilateral asymmetry, landmarking also plays a major role in appropriate matching of structures of the two breasts. For the case of the pectoral muscle, the algorithm proposed by Ferrari et al. [49] was applied and novel constrains have been designed. Two novel approaches have been presented and validated for extraction of the breast skin-line and detection of the nipple.

The proposed procedure for the estimation of the breast skin-line is based upon multi-directional Gabor filtering. The method includes an adaptive values-of-interest transformation, extraction of the skin-air ribbon by Otsu's thresholding method and the Euclidean distance transform, Gabor filtering with 18 real kernels, and a step for suppression of false edge points using the magnitude and phase responses of the filters, followed by an edge-linking algorithm. On a test set of 361 images from different acquisition modalities, SFMs and FFDMs, the average Hausdorff and polyline distances obtained were 2.85 mm and 0.84 mm, respectively, with reference to the ground-truth boundaries provided by an expert radiologist. When compared with the results obtained by other state-of-the-art methods on the same set of images and with respect to the same ground-truth boundaries, the proposed method outperformed the other approaches. The results have demonstrated the effectiveness and robustness of the proposed algorithm.

A novel Hessian-based method was applied to locate automatically the nipple. The method includes detection of a plausible nipple/retroareolar area in a mammogram using geometrical constraints, analysis of the GVF by mean and Gaussian curvature measurements, and local shape-based conditions. The proposed procedure was tested on 566 mammographic images. A radiologist independently marked the centers of the nipples for evaluation of the results. The method was directly compared with two other techniques for detection of the nipple and evaluated in terms of the Euclidean distance between the automatically detected position and the center of the nipple as identified by the radiologist. The average error obtained was 6.7 mm. Only two out of the 566 detected nipple positions (0.35%) had an error larger than 50 mm. Results and comparative analysis indicate that the proposed method outperforms other algorithms presented in the literature and can be used to identify accurately the nipple on various types of mammographic images.

Based on the state of the art in the field, two main lines of investigation have been addressed:

1. the development and validation of an automatic system for the identification of bilateral asymmetry as an early sign of breast cancer; and

2. the development and validation of an integrated CADe/CADx system that can detect masses and regions of architectural distortion on mammograms and can provide, in addition, their characterization in terms of malignancy.

For the task of detecting bilateral asymmetry, following radiologists' criteria in the interpretation of mammograms, bilateral masking procedures have been designed to match the regions of the breast to be compared by means of the automatically detected anatomical landmarks. It was hypothesized that quantification of structural similarity or dissimilarity between paired mammographic regions could be effective in detecting asymmetric signs of breast cancer. Changes in structural information of the extracted regions were investigated using three different approaches. The first approach is based on a new index of phase similarity that measures the angular covariance between rose diagrams related to the phase and magnitude responses of multidirectional Gabor filters. Analysis of spatial correlation with respect to the nipple position was performed as a second approach to quantify differences in the spatial distribution of pixel values. The third approach uses spherical semivariogram descriptors and correlation-based structural similarity indices in the spatial and complex wavelet domains. The spatial distribution of grayscale values as well as of the magnitude and phase responses of multidirectional Gabor filters were used to represent the structure of mammographic density and of the directional components of breast tissue patterns, respectively.

A total of 188 mammograms from the DDSM and mini-MIAS databases, consisting of 47 asymmetric cases and 47 normal cases, were analyzed. For the combined dataset of mammograms, areas under the ROC curves of 0.83, 0.77, and 0.87 were obtained, respectively, with LDA, BQDA, and ANN-RBF, using the features selected by SWR and leave-one-patient-out cross-validation. Two-view analysis provided accuracy up to 0.94, with sensitivity and specificity of 1 and 0.88, respectively. The performance achieved demonstrates the effectiveness of combining landmarking, automatic masking procedures, Gabor filters, spherical semivariograms, and measures of similarity in both spatial and wavelet domains, to detect bilateral structural asymmetry. The study also demonstrates the feasibility of accurately quantifying bilateral differences between mammograms for detection of subtle signs of pathological origin. Another possible application may be retrospective comparison of current and prior mammograms of the same patient to determine whether any changes occurred over time. The development of CAD techniques for detection of bilateral asymmetry is expected to improve the performance of interpretation techniques for mammography and breast cancer.

As part of an integrated CADe/CADx system, an additional contribution of the present work has been the design of contour-independent features for classification of breast masses as malignant tumors or benign lesions. This aspect is important due to the presence of tumors with obscured or ill-defined margins for which the existing approaches, based on accurate segmentation of the lesions, are prone to fail. The new features quantify nonstationarity and spatial dependence of pixel values in the angular and radial directions. Using an ANN-RBF to predict the diagnosis of 120 benign masses and 26 malignant tumors in a database of FFDMs, an area under the ROC curve of 0.99 ± 0.01 was obtained using randomly selected centroidal points and an expected size of the masses. Results indicate that the use of the proposed contour-independent features can be an effective approach for automatic classification of mammographic lesions.

The final objective achieved in our study consisted of a multistage approach to mammographic detection and characterization of breast cancer. The ultimate goal was to discriminate malignant tumors from benign lesions and normal parenchymal tissue in a realistic scenario of lesion candidates automatically detected in mammograms. Local analysis of the Gaussian curvature and of the phase response of multidirectional Gabor filters was performed for identification of suspicious focal areas. The detection and classification of lesions were performed in series, respectively, via a differential approach to analysis of the tissue surrounding the candidates and via quantification of nonstationarity and spatial dependence of pixel values within circular and annular ROIs. A new unified 3D FROC framework has been introduced for global analysis of the two binary categorization problems in series. The system was tested on a total of 2105 FFDMs and SFMs from three different datasets, including abnormal mammograms with 560 malignant tumors and 639 benign lesions, masses, or architectural distortion, and 1010 normal mammograms. For sensitivity of detection of malignant tumors in the range of 0.70–0.81, the range of falsely detected malignant tumors per image was [0.82–3.47], with a series of two stages of classification, including SWR for selection of features, FLDA, and two-fold cross-validation.

Integrated systems for automatic detection and diagnosis of breast cancer can facilitate accurate interpretation of mammograms and efficient treatment. The results reported in this study on a large and diversified dataset indicate that CAD of malignant tumors is possible even without accurate extraction of their contours.

Future work is desirable to improve further the diagnostic performance of CAD systems for mammography and to reach increased performance levels. Possible trends of research in the area of computerized analysis of mammograms include quantitative assessment of near-term risk for having or developing breast cancer, development of individualized and optimal risk-based screening plans, and retrospective comparison of current and prior mammograms of the same patient to detect and analyze any changes that might have occurred over time. Additional work is in progress to locate the position of asymmetric findings on mammograms and to improve the obtained classification performance by means of an ensemble of classifiers.

The presented CAD techniques and their application in clinical practice are expected to help radiologists in improving the life expectancy of breast cancer patients via early detection and accurate diagnosis.

References

[1] American Cancer Society. Breast density and your mammogram report, 2015. http://www.cancer.org/ 130

[2] P. Autier, M. Boniol, R. Middleton, J. Doré, C. Héry, T. Zheng, and A. Gavin. Advanced breast cancer incidence following population-based mammographic screening. *Ann. Oncol.*, 22:1726–1735, 2011. DOI: 10.1093/annonc/mdq633. 9

[3] S. R. Aylward, B. Hemminger, and E. Pisano. Mixture modeling for digital mammogram display and analysis. In *Digital Mammography (IWDM 1998) in Computational Imaging and Vision*, pages 305–312. Springer, 1998. DOI: 10.1007/978-94-011-5318-8_51. 30

[4] F. Ayres and R. Rangayyan. Design and performance analysis of oriented feature detectors. *J. Electron. Imaging*, 16(2):12, April 2007. article number 023007. DOI: 10.1117/1.2728751. 24, 25, 27

[5] M. Bazzocchi, F. Mazzarella, C. D. Frate, R. Girometti, and C. Zuiani. CAD systems for mammography: A real opportunity? A review of the literature. *Radiol. Med.*, 112(3): 329–353, 2007. DOI: 10.1007/s11547-007-0145-5. 5

[6] T. Beck, T. Housh, J. Weir, J. Cramer, V. Vardaxis, G. Johnson, J. Coburn, M. Malek, and M. Mielke. An examination of the runs test, reverse arrangements test, and modified reverse arrangements test for assessing surface EMG signal stationarity. *J. Neurosci. Meth.*, 156(1–2):242–248, Sep. 2006. DOI: 10.1016/j.jneumeth.2006.03.011. 106

[7] J. Bendat and A. Piersol. *Random Data: Analysis and Measurement Procedures*. John Wiley & Sons, 1986. DOI: 10.1002/9781118032428. 116, 121

[8] S. Berger, B. Curcio, J. Gershoncohen, and H. Isard. Mammographic localization of unsuspected breast cancer. *Am. J. Roentgenol. Radium Ther. Nucl. Med.*, 96:1046–1052, 1966. DOI: 10.2214/ajr.96.4.1046. 130

[9] J. Bezdec. *Pattern Recognition with Fuzzy Objective Function Algorithms*. Plenum Press, New York, 1981. 130

[10] U. Bick, M. Giger, R. Schmidt, R. Nishikawa, D. Wolverton, and K. Doi. Automated segmentation of digitized mammograms. *Acad. Radiol.*, 2(1):1–9, 1995. DOI: 10.1016/s1076-6332(05)80239-9. 31

[11] R. Birdwell, P. Bandodkar, and D. Ikeda. Computer-aided detection with screening mammography in a university hospital setting. *Radiology*, 236(2):451–457, 2005. DOI: 10.1016/s0098-1672(08)70300-x. 12, 13

[12] N. Boyd, H. Guo, L. Martin, L. Sun, J. Stone, E. Fishell, R. Jong, A. Chiarelli, S. Minkin, and M. Yaffe. Mammographic density and the risk and detection of breast cancer. *N. Engl. J. Med.*, 356(3):227–236, 2007. DOI: 10.1056/nejmoa062790. 76

[13] G. A. Bradshaw and B. Finlay. Natural symmetry. *Nature*, 7039(435):149–149, 2005. DOI: 10.1038/435149a. 75

[14] J. Bronzino. *Medical Devices and Systems*, 3rd ed. Taylor and Francis, 2006. DOI: 10.1201/9781420003864. 2

[15] H. Burrell, A. Evans, A. Wilson, and S. Pinder. False-negative breast screening assessment. What lessons can we learn? *Clin. Radiol.*, 56(5):385–388, 2001. DOI: 10.1053/crad.2001.0662. 7, 76

[16] P. Casti, A. Mencattini, and M. Salmeri. Characterization of the breast region for computer assisted Tabár masking of paired mammographic images. In *IEEE Computer-based Medical Systems (CBMS), 25th International Symposium on*, pages 1–6, 2012. DOI: 10.1109/cbms.2012.6266363. 80

[17] P. Casti, A. Mencattini, M. Salmeri, and M. Pepe. Fibroglandular tissue segmentation using a Gaussian mixture model and a novel kernel merging technique. In *26th International Congress and Exhibition: Computer Assisted Radiology and Surgery*, 7(1):S307–S365, Pisa, Italy, 2012. 29

[18] P. Casti, A. Mencattini, M. Salmeri, A. Ancona, F. Mangieri, M. Pepe, and R. Rangayyan. Automatic detection of the nipple in screen-film and full-field digital mammograms using a novel Hessian-based method. *J. Digit. Imag.*, 26(5):948–957, 2013. DOI: 10.1007/s10278-013-9587-6. 42, 52

[19] P. Casti, A. Mencattini, M. Salmeri, A. Ancona, F. Mangieri, M. Pepe, and R. Rangayyan. Estimation of the breast skin-line in mammograms using multidirectional Gabor filters. *Comput. Biol. Med.*, 43(11):1870–1881, 2013. DOI: 10.1016/j.compbiomed.2013.09.001. 54, 55, 59, 60, 66, 69, 70, 72

[20] P. Casti, A. Mencattini, M. Salmeri, A. Ancona, F. Mangieri, and R. Rangayyan. Measures of radial correlation and trend for classification of breast masses in mammograms. In *Conf. Proc. IEEE Eng. Med. Biol. Soc. (EMBS)*, Osaka, Japan, Jul. 3–7, 2013. DOI: 10.1109/embc.2013.6611041. 106, 108, 109, 111

[21] P. Casti, A. Mencattini, M. Salmeri, A. Ancona, F. Mangieri, and R. Rangayyan. Design and analysis of contour-independent features for classification of mammographic lesions. In *Conf. Proc. IEEE e-Health and Bioeng. (EHB)*, Iasi, Romania, 21–23 Nov. 2013. DOI: 10.1109/ehb.2013.6707401. 80, 102, 106, 115, 116

[22] P. Casti, A. Mencattini, M. Salmeri, A. Ancona, F. Mangieri, and R. Rangayyan. Masking procedures and measures of angular similarity for detection of bilateral asymmetry in mammograms. In *Conf. Proc. IEEE e-Health and Bioeng. (EHB)*, Iasi, Romania, 21–23 Nov. 2013. DOI: 10.1109/ehb.2013.6707258. 79, 85

[23] P. Casti, A. Mencattini, M. Salmeri, A. Ancona, F. Mangieri, and R. Rangayyan. Development and validation of a fully automated system for detection and diagnosis of mammographic lesions. In *Conf. Proc. IEEE Eng. Med. Biol. Soc. (EMBS)*, Chicago, Aug. 26–30, 2014. DOI: 10.1109/embc.2014.6944665. 109, 116, 120

[24] P. Casti, A. Mencattini, M. Salmeri, and R. Rangayyan. Semivariogram analysis and spherical modeling to detect structural bilateral asymmetry in mammograms. *Int. J. CARS*, 9(1):S231–S234, 2014. 79, 80, 92, 132

[25] P. Casti, A. Mencattini, M. Salmeri, and R. Rangayyan. Spatial correlation analysis of mammograms for detection of asymmetric findings. In *Breast Imaging (IWDM 2014) in Lecture Notes in Computer Science*, pages 558–564. Springer, 2014. DOI: 10.1007/978-3-319-07887-8_78. 79, 80, 102, 141

[26] P. Casti, A. Mencattini, M. Salmeri, A. Ancona, F. Mangeri, M. Pepe, and R. Rangayyan. Contour-independent detection and classification of mammographic lesions. *Biomed. Signal Process. Control*, 25:165–177, 2015. DOI: 10.1016/j.bspc.2015.11.010. 109, 120, 122, 124, 126, 127, 128, 133, 136, 137, 138

[27] P. Casti, A. Mencattini, M. Salmeri, and R. Rangayyan. Analysis of structural similarity in mammograms for detection of bilateral asymmetry. *IEEE Trans. Med. Imag.*, 34(2): 662–671, 2015. DOI: 10.1109/tmi.2014.2365436. 79, 80, 81, 82, 83, 93, 95, 96, 101

[28] P. Casti, A. Mencattini, M. Salmeri, A. Ancona, M. Lorusso, M. Pepe, C. D. Natale, and E. Martinelli. Towards localization of malignant sites of asymmetry across bilateral mammograms. *Comput. Meth. Progr. Biomed.*, 140:11–18, 2017. DOI: 10.1016/j.cmpb.2016.11.010. 103

[29] T. Chan and L. Vese. Active contours without edges. *IEEE Trans. Image Process.*, 10(1): 266–277, 2001. DOI: 10.1109/83.902291. 43

[30] R. Chandrasekhar and Y. Attikiouzel. A simple method for automatically locating the nipple on mammograms. *IEEE Trans. Med. Imag.*, 16(5):483–494, 1197. DOI: 10.1109/42.640738. 30, 53

[31] H. Cheng, X. Cai, X. Chen, L. Hu, and X. Lou. Computer-aided detection and classification of microcalcifications in mammograms: A survey. *Pattern Recogn.*, 36(12):2967–2991, 2003. DOI: 10.1016/s0031-3203(03)00192-4. 119

[32] H. Cheng, X. Shi, R. Min, L. Hu, X. Cai, and H. Du. Approaches for automated detection and classification of masses in mammograms. *Pattern Recogn.*, 39(4):646–668, 2006. DOI: 10.1016/j.patcog.2005.07.006. 120

[33] Y. Chun and D. Griffith. *Spatial statistics and geostatistics: Theory and applications for geographic information science and technology.* SAGE, 2013. 86

[34] S. Ciatto, M. D. Turco, P. Burke, C. Visioli, E. Paci, and M. Zappa. Comparison of standard and double reading and computer-aided detection (CAD) of interval cancers at prior negative screening mammograms: Blind review. *Br. J. Cancer*, 89:1645–1649, 2003. DOI: 10.1038/sj.bjc.6601356. 12, 13

[35] F. H. J. Cornish. Symmetries and curvature structure in general relativity. *Class. Quant. Gravity*, 21(21):5019–5020, 2004. 75

[36] D. Davies and D. Dance. Automatic computer detection of clustered calcifications in digital mammograms. *Phys. Med. Biol.*, 35(8):1111–1118, 1990. DOI: 10.1088/0031-9155/35/8/007. 31

[37] K. Doi. Computer-aided diagnosis in medical imaging: Historical review, current status and future potential. *J. Comp. Med. Imag. Graph.*, 31(4–5):198–211, 2007. DOI: 10.1016/j.compmedimag.2007.02.002. 77

[38] A. Domínguez and A. Nandi. Toward breast cancer diagnosis based on automated segmentation of masses in mammograms. *Pattern Recogn.*, 42:1138–1148, 2009. DOI: 10.1016/j.patcog.2008.08.006. 105, 106, 117, 120

[39] C. D'Orsi, L. Bassett, W. Berg, S. Feig, V. Jackson, and D. Kopans. *BI-RADS: Mammography*, 4th ed. American College of Radiology, 2003. 3, 4, 5, 7, 123, 124, 125, 135

[40] E. Dowling, C. Klabunde, J. Patnick, and R. Ballard-Barbash. Breast and cervical cancer screening programme implementation in 16 countries. *J. Med. Screen.*, 17:139–146, 2010. DOI: 10.1258/jms.2010.010033. 1

[41] N. Draper and H. Smith. *Applied Regression Analysis.* Wiley-Interscience, 1998. DOI: 10.1002/9781118625590. 94

[42] R. Duda, P. Hart, and D. Stork. *Pattern Classification*, 2nd ed. Wiley-Interscience, 2001. DOI: 10.1007/s00357-007-0015-9. 87, 94, 110

[43] M. Elter and A. Horsch. CADx of mammographic masses and clustered microcalcifications: A review. *Med. Phys.*, 36(6):2052–2068, 2009. DOI: 10.1118/1.3121511. 105, 106, 120

[44] N. Eltonsy, G. Tourassi, and A. Elmaghraby. A concentric morphology model for the detection of masses in mammography. *IEEE Trans. Med. Imag.*, 26(6):880–889, 2007. DOI: 10.1109/tmi.2007.895460. 124, 140

[45] D. Ericeira, A. Silva, A. de Paiva, and M. Gattass. Detection of masses based on asymmetric regions of digital bilateral mammograms using spatial description with variogram and cross-variogram functions. *Comput. Biol. Med.*, 43(8):987–999, 2013. DOI: 10.1016/j.compbiomed.2013.04.019. 77, 90

[46] J. Fenton, S. Taplin, P. Carney, L. Abraham, E. Sickles, C. D'Orsi, E. Berns, G. Cutter, E. Hendrick, W. Barlow, and J. Elmore. Influence of computer-aided detection on performance of screening mammography. *N. Engl. J. Med.*, 356:1399–1409, 2007. DOI: 10.1056/nejmoa066099. 13

[47] R. Ferrari, R. Rangayyan, J. Desautels, and A. Frère. Segmentation of mammograms: Identification of the skin—air boundary, pectoral muscle, and fibro-glandular disc. In *Proc. 5th Int. Workshop Digital Mammography*, pages 573–579, 2000. 29

[48] R. Ferrari, R. Rangayyan, J. Desautels, and A. Frère. Analysis of asymmetry in mammograms via directional filtering with Gabor wavelets. *IEEE Trans. Med. Imag.*, 20(9): 953–964, 2001. DOI: 10.1109/42.952732. 77, 90, 102

[49] R. Ferrari, R. Rangayyan, J. Desautels, R. Borges, and A. Frère. Automatic identification of the pectoral muscle in mammograms. *IEEE Trans. Med. Imag.*, 23(2):232–245, 2004. DOI: 10.1109/tmi.2003.823062. 29, 30, 33, 35, 36, 57, 143

[50] R. Ferrari, R. Rangayyan, J. Desautels, R. Borges, and A. Frère. Identification of the breast boundary in mammograms using active contour models. *Med. Biol. Eng. Comput.*, 42(2): 201–208, 2004. DOI: 10.1007/bf02344632. 21, 31, 32, 56, 63, 64, 65, 68

[51] F. Foca, S. Mancini, L. Bucchi, D. Puliti, M. Zappa, C. Naldoni, F. Falcini, M. Gambino, S. Piffer, M. Gonzalez, F. Stracci, M. Zorzi, E. Paci, and the IMPACT Working Group. Decreasing incidence of late-stage breast cancer after the introduction of organized mammography screening in Italy. *Cancer*, 24(2):280–285, 2013. DOI: 10.1002/cncr.28014. 1

[52] T. Freer and M. Ulissey. Screening mammography with computer-aided detection: Prospective study of 12,860 patients in a community breast center. *Radiology*, 220(3): 781–786, 2001. DOI: 10.1148/radiol.2203001282. 12, 13

[53] D. Gabor. Theory of communications. *J. Inst. Electr. Eng.*, 3(93):429–457, Feb. 1967. DOI: 10.1049/ji-3-2.1946.0074. 108, 114

[54] F. Gilbert, S. Astley, M. Gillan, O. Agbaje, M. Wallis, J. James, C. Boggis, and S. Duffy. Single reading with computer-aided detection for screening mammography. *N. Engl. J. Med.*, 359(16):1675–1684, 2008. DOI: 10.1056/nejmoa0803545. 9, 12, 13, 76

[55] R. Gonzalez and R. Woods. *Digital Image Processing*, 3rd ed. Prentice Hall, 2008. 62

[56] A. Gray, E. Abbena, and S. Salamon. *Modern Differential Geometry of Curves and Surfaces with Mathematica*. Chapman & Hall/CRC, 2006. 129

[57] D. J. Gross. The role of symmetry in fundamental physics. *Proc. of the National Academy of Sciences*, 93(25):14256–14259, 1996. DOI: 10.1073/pnas.93.25.14256. 75

[58] A. Hapfelmeier and A. Horsch. Image feature evaluation in two new mammography CAD prototypes. *Int. J. Comput. Assist. Radiol. Surg.*, 6:721–735, 2011. DOI: 10.1007/s11548-011-0549-5. 105, 106, 117

[59] J. Harvey, L. Fajardo, and C. Innis. Previous mammograms in patients with impalpable breast carcinoma: Retrospective vs. blinded interpretation. *Am. J. Roentgenol.*, 161(6): 1167–1172, 1993. DOI: 10.2214/ajr.161.6.8249720. 7, 76

[60] S. Haykin. *Neural Networks: A Comprehensive Foundation*. Prenctice Hall, 1999. 94, 110, 116

[61] M. Heath, K. Bowyer, D. Kopans, R. Moore, and W. Kegelmeyer. The Digital Database for Screening Mammography. In *Proc. 5th International Workshop on Digital Mammography*, pages 212–218. Medical Physics Publishing, 2001. 4, 5, 6, 8, 20, 21, 31, 33, 81, 90, 93, 95, 121, 123

[62] S. Hofvind, P. Skaane, J. Elmore, S. Sebuodegard, S. Hoff, and C.Lee. Mammographic performance in a population-based screening program: Before, during, and after the transition from screen-film to full-field digital mammography. *Radiology*, 272(1):52–62, 2014. DOI: 10.1148/radiol.14131502. 2

[63] J. Iglesias and N. Karssemeijer. Robust initial detection of landmarks in film-screen mammograms using multiple ffdm atlases. *IEEE Trans. Med. Imag.*, 28(11):1815–1824, 2009. DOI: 10.1109/tmi.2009.2025036. 30, 53

[64] International Agency for Research on Cancer (AIRC). GLOBOCAN: Estimated cancer incidence mortality and prevalence worldwide in 2012. http://www.iarc.fr/ accessed January 2015. 1

[65] E. Isaaks and R. Srivastava. *An Introduction to Applied Geostatistics*. Oxford University Press, 1989. 90

[66] Italian National Institute for Statistics (ISTAT). Leading causes of death in Italy year 2012. http://www.istat.it/en/, 2014. accessed January 2015. 1

[67] M. Karnan and K. Thangavel. Automatic detection of the breast border and nipple position on digital mammograms using genetic algorithm for asymmetry approach to detection of microcalcifications. *Comput. Meth. Progr. Biomed.*, 87(1):12–20, 2007. DOI: 10.1016/j.cmpb.2007.04.007. 30, 32, 77

[68] N. Karssemeijer. Automated classification of parenchymal patterns in mammograms. *Phys. Med. Biol.*, 43(2):365–378, 1998. DOI: 10.1088/0031-9155/43/2/011. 29, 35, 36, 77

[69] N. Karssemeijer. Detection of masses in mammograms. In *Image-Processing Techniques for Tumor Detection*, pages 187–212. CRC Press, 2002. DOI: 10.1201/9780203909355.ch8. 4

[70] N. Karssemeijer and G. te Brake. Detection of stellate distortions in mammograms. *IEEE Trans. Med. Imag.*, 15(5):611–619, 1996. DOI: 10.1109/42.538938. 124, 140

[71] A. Khademi, D. Hosseinzadeh, A. Venetsanopoulos, and A. Moody. Nonparametric statistical tests for exploration of correlation and nonstationarity in images. In *Proc. 16th Int. Conf. on Dig. Sign. Proc.*, pages 1–6, Santorini-Hellas, Greece, July 2009. DOI: 10.1109/icdsp.2009.5201186. 106

[72] S. Kinoshita, P. de Azevedo-Marques, R. P. J. J. Rodrigues, and R. Rangayyan. Content-based retrieval of mammograms using visual features related to breast density patterns. *J. Digit. Imag.*, 20(2):172–190, 2007. DOI: 10.1007/s10278-007-9004-0. 39, 42

[73] S. Kinoshita, P. Azevedo-Marques, R. Pereira, J. Rodrigues, and R. Rangayyan. Radon-domain detection of the nipple and the pectoral muscle in mammograms. *J. Digit. Imag.*, 21(1):37–49, 2008. DOI: 10.1007/s10278-007-9035-6. 30, 50, 51, 52

[74] G. Klanderman and W. Rucklidge. Comparing images using the Hausdorff distance. *IEEE Trans. Pattern Anal. Machine Intell.*, 15(9):850–863, 1993. DOI: 10.1109/34.232073. 64

[75] C. P. Klingenberg. Analyzing fluctuating asymmetry with geometric morphometrics: concepts, methods, and applications. *Symmetry*, 2(7):843–934, 2015. DOI: 10.3390/sym7020843. 75

[76] J. Ko, M. Nicholas, J. Mendel, and P. Slanetz. Prospective assessment of computer-aided detection in interpretation of screening mammography. *Am. J. Roentgenol.*, 187(6):1483–1491, 2006. DOI: 10.2214/ajr.05.1582. 12, 13

[77] G. Kom, A. Tiedeu, and M. Kom. Automated detection of masses in mammograms by local adaptive thresholding. *Comput. Biol. Med.*, 37(1):37–48, 2007. DOI: 10.1016/j.compbiomed.2005.12.004. 124, 140

[78] V. Kumar, A. Abbas, N. Fausto, and J. Aster. *Pathological Basis of Disease*, 8th ed. Saunders Elsevier, 2009. 130

[79] P. Kus and I. Karagoz. Fully automated gradient based breast boundary detection for digitized X-ray mammograms. *Comput. Biol. Med.*, 42(1):75–82, 2012. DOI: 10.1016/j.compbiomed.2011.10.011. 32, 56, 63, 64, 65

[80] S. Lai, X. Li, and W. Bischof. On techniques for detecting circumscribed masses in mammograms. *IEEE Trans. Med. Imag.*, 8(4):377–386, 1989. DOI: 10.1109/42.41491. 124, 140

[81] T. Lau and W. Bischof. Automated detection of breast tumors using the asymmetry approach. *Comput. Biomed. Res.*, 24:273–295, 1991. DOI: 10.1016/0010-4809(91)90049-3. 77, 102

[82] I. Leconte, C. Feger, C. Galant, M. Berliére, B. Berg, W. D'Hoore, and B. Maldague. Mammography and subsequent whole-breast sonography of nonpalpable breast cancers: the importance of radiologic breast density. *Am. J. Roentgenol.*, 180(6):1675–1679, 2003. DOI: 10.2214/ajr.180.6.1801675. 7

[83] P. Legendre and M. Fortin. Spatial pattern and ecological analysis. *Vegetatio*, 80(2):107–138, 1989. DOI: 10.1007/bf00048036. 121

[84] C. Lerman, K. Kash, and M. Stefanek. Younger women at increased risk for breast cancer: Perceived risk, psychological well-being, and surveillance behavior. *J. Natl. Cancer Inst. Monogr.*, 16(16):171–176, 1994. ISSN 1052-6773. http://europepmc.org/abstract/MED/7999461 1

[85] W. Ludwig. *Das Rechts-Links Problem im Tierreich und beim Menschen: mit einem Anhang Rechts-links-Merkmale der Pflanzen.* Springer-Verlag, 1932. DOI: 10.1007/978-3-662-29196-2. 75

[86] G. Magna, S. V. Jayaraman, P. Casti, A. Mencattini, C. D. Natale, and E. Martinelli. Adaptive classification model based on artificial immune system for breast cancer detection. In *AISEM Annual Conference, XVIII*, pages 1–4. IEEE, 2015. DOI: 10.1109/aisem.2015.7066842. 103

[87] G. Magna, P. Casti, S. Jayaraman, M. Salmeri, A. Mencattini, E. Martinelli, and C. C. Di Natale. Identification of mammography anomalies for breast cancer detection by an ensemble of classification models based on artificial immune system. *Knowledge-Based Syst.*, 101:60–70, 2016. DOI: 10.1016/j.knosys.2016.02.019. 103

[88] A. Majid, E. de Paredes, R. Doherty, N. Sharma, and X. Salvador. Missed breast carcinoma: Pitfalls and pearls. *Radiographics*, 23:881–895, 2003. DOI: 10.1148/rg.234025083. 7, 9, 76, 105

[89] B. S. Manjunath and W. Y. Ma. Texture features for browsing and retrieval of image data. *IEEE Trans. Pattern Anal. Mach. Intell.*, 18:837–842, 1996. DOI: 10.1109/34.531803. 30

[90] N. Mantel. The detection of disease clustering and a generalized regression approach. *Cancer Res.*, 27(2):209–220, Feb. 1967. 106

[91] R. Martí, A. Oliver, D. Raba, and J. Freixenet. Breast skin-line segmentation using contour growing. In *Pattern Recognition and Image Analysis*, pages 564–571. Springer Berlin, 2007. DOI: 10.1007/978-3-540-72849-8_71. 31

[92] C. Maurer, Jr., R. Qi, and V. Raghavan. A linear time algorithm for computing exact Euclidean distance transforms of binary images in arbitrary dimensions. *IEEE Trans. Pattern Anal. Mach. Intell.*, 25(2):265–270, 2003. DOI: 10.1109/tpami.2003.1177156. 54

[93] A. Mencattini and M. Salmeri. Breast masses detection using phase portrait analysis and fuzzy inference systems. *Int. J. Comput. Assist. Radiol. Surg.*, 7(4):573–583, 2011. DOI: 10.1007/s11548-011-0659-0. 32, 46, 102, 141

[94] A. Mencattini, M. Salmeri, and P. Casti. Bilateral asymmetry identification for the early detection of breast cancer. In *Medical Measurements and Applications Proceedings (MeMeA), IEEE International Workshop on*, pages 613–618, Bari, Italy, 2011. DOI: 10.1109/memea.2011.5966746. 29

[95] A. Mencattini, M. Salmeri, P. Casti, M. Pepe, and A. Ancona. Local active contour models and Gabor wavelets for an optimal breast region segmentation. In *26th International Congress and Exhibition: Computer Assisted Radiology and Surgery*, 7(1):S256–S257, Pisa, Italy, 2012. Int. J. CARS. 32, 33, 37, 41

[96] A. Méndez, P. Tahoces, M. Lado, M. Souto, J. Correa, and J. Vidal. Automatic detection of breast border and nipple in digital mammograms. *Comp. Meth. Progr. Biomed.*, 49(3): 253–262, 1996. DOI: 10.1016/0169-2607(96)01724-5. 31, 53

[97] A. Miller, C. Wall, C. Baines, P. Sun, T. To, and S. Narod. Twenty five year follow-up for breast cancer incidence and mortality of the Canadian National Breast Screening Study: Randomised screening trial. *Br. Med. J.*, 348:g366, 2014. DOI: 10.1136/bmj.g366. 9

[98] P. Miller and S. Astley. Automated detection of mammographic asymmetry using anatomical features. *Int. J. Pattern Recogn. Artif. Intell.*, 7(6):1461–1476, 1993. DOI: 10.1142/s0218001493000716. 77, 90, 102

156 REFERENCES

[99] A. P. Møller and J. P. Swaddle. *Asymmetry, Developmental Stability and Evolution*. Oxford University Press, UK, 1997. 75

[100] P. Moran. Notes on continuous stochastic phenomena. *Biometrika*, 37(1/2):17–23, 1950. DOI: 10.1093/biomet/37.1-2.17. 86

[101] M. Morton, D. Whaley, K. Brandt, and K. Amrami. Screening mammograms: Interpretation with computer-aided detection—prospective evaluation. *Radiology*, 239(2):375–383, 2006. DOI: 10.1148/radiol.2392042121. 12, 13

[102] N. Mudigonda, R. Rangayyan, and J. Desautels. Detection of breast masses in mammograms by density slicing and texture flow-field analysis. *IEEE Trans. Med. Imag.*, 20(12): 1215–1227, 2001. DOI: 10.1109/42.974917. 105, 117, 120, 132

[103] N. E. M. A. (NEMA). Digital Imaging and Communication in Medicine (DICOM). Rossyln, VA, 2011. 19, 107

[104] T. Ojala, J. Näppi, and O. Nevalainen. Accurate segmentation of the breast region from digitized mammograms. *Computer. Med. Imag. Graph.*, 25(1):47–59, 2001. DOI: 10.1016/s0895-6111(00)00036-7. 31

[105] A. Oliver, J. Freixenet, J. Martí, E. Pérez, J. Pont, E. Denton, and R. Zwiggelaar. A review of automatic mass detection and segmentation in mammographic immages. *Med. Image Anal.*, 14(2):87–110, 2010. DOI: 10.1016/j.media.2009.12.005. 119, 120, 124, 140, 141

[106] M. Oliver. Determining the spatial scale of variation in environmental properties using the variogram. In N. Tate and P. Atkinson, Eds., *Modelling Scale in Geographical Information Science*, pages 193–219. John Wiley & Sons, 2001. 90

[107] N. Otsu. A threshold selection method from gray-level histograms. *IEEE Trans. Syst., Man, Cybern.*, 9(1):62–66, 1979. DOI: 10.1109/tsmc.1979.4310076. 31, 54, 61

[108] J. Padayachee, M. Alport, and W. Rae. Identification of the breast edge using areas enclosed by iso-intensity contours. *Computer. Med. Imag. Graph.*, 31(6):390–400, 2007. DOI: 10.1016/j.compmedimag.2007.02.019. 32

[109] A. Palmer and C. Strobeck. Fluctuating asymmetry: measurement, analysis, patterns. *Annu. Rev. Ecol. Evol. Syst.*, 17:391–421, 1986. DOI: 10.1146/annurev.es.17.110186.002135. 75

[110] A. R. Palmer. Symmetry breaking and the evolution of development. *Science*, 5697(306): 828–833, 2004. DOI: 10.1126/science.1103707. 75

[111] S. Paquerault, N. Petrick, H. Chan, B. Sahiner, and M. Helvie. Improvement of computerized mass detection on mammograms: Fusion of two-view information. *Med. Phys.*, 29(2):238–247, 2002. DOI: 10.1118/1.1446098. 39

[112] N. Perry, M. Broeders, C. de Wolf, S. Törnberg, and J. Schouten. *European Guidelines for Quality Assurance in Mammography Screening* 3rd ed. European Commission, 2006. DOI: 10.1093/annonc/mdm481. 1

[113] W. Peterson, T. Birdsall, and W. Fox. The theory of signal detectability. *IEEE Trans. Inform. Theor.*, 4(4):171–212, 1954. DOI: 10.1109/tit.1954.1057460. 15

[114] N. Petrick, B. Sahiner, H. Chan, M. Helvie, S. Paquerault, and L. Hadjiisky. Breast cancer detection: Evaluation of a mass-detection algorithm for computer-aided diagnosis— experience in 263 patients. *Radiology*, 224(1):217–224, 2002. DOI: 10.1148/radiol.2241011062. 124, 140

[115] J. Picard. History of mammography. *Bull. Acad. Natl. Med.*, 182(8):1613–1620, 1998. 2

[116] W. Polakowski, D. Cournoyer, S. Rogers, M. DeSimio, D. Ruck, J. Hoffmeister, and R. Raines. Computer-aided breast cancer detection and diagnosis of masses using difference of Gaussians and derivative-based feature saliency. *IEEE Trans. Med. Imag.*, 16(6): 811–819, 1997. DOI: 10.1109/42.650877. 120, 124, 140

[117] J. Portilla and E. Simoncelli. A parametric texture model based on joint statistics of complex wavelet coefficients. *Int. J. Comput. Vision*, 40(1):49–71, 2000. DOI: 10.1023/A:1026553619983. 92, 94

[118] D. Raba, A. Oliver, J. Martí, M. Peracaula, and J. Espunya. Breast segmentation with pectoral muscle suppression on digital mammograms. In S. B. Heidelberg., Ed., *Iberian Conference on Pattern Recognition and Image Analysis*, pages 471–478, 2005. 31

[119] R. Ramos-Pollán, M. Guevara-López, C. Suárez-Ortega, G. Díaz-Herrero, J. Franco-Valiente, M. R. del Solar, N. G. de Posada, M. Vaz, J. Loureiro, and I. Ramos. Discovering mammography-based machine learning classifiers for breast cancer diagnosis. *J. Med. Syst.*, 36:2259–2269, 2012. DOI: 10.1007/s10916-011-9693-2. 105, 106

[120] F. Ramsey and D. Schafer. *The Statistical Sleuth: A Course in Methods of Data Analysis*. Duxbury Press, Belmont, CA, 1997. 110

[121] R. Rangayyan. *Biomedical Image Analysis*. CRC Press, 2005. DOI: 10.1201/9780203492543. 15, 77

[122] R. Rangayyan and T. Nguyen. Fractal analysis of contours of breast masses in mammograms. *J. Digit. Imag.*, 4:223–237, 2007. DOI: 10.1007/s10278-006-0860-9. 105, 111

[123] R. Rangayyan, N. El-Faramawy, J. Desautels, and O. Alim. Measures of acutance and shape for classification of breast tumors. *IEEE Trans. Med. Imag.*, 16:799–810, 1997. DOI: 10.1109/42.650876. 111

[124] R. Rangayyan, R. Ferrari, and A. Frère. Analysis of bilateral asymmetry in mammograms using directional, morphological, and density features. *J. Electron. Imag.*, 16(1):12, 2007. article number 013003. DOI: 10.1117/1.2712461. 39, 77, 90, 102, 119

[125] R. Rangayyan, S. Banik, and J. Desautels. Computer-aided detection of architectural distortion in prior mammograms of interval cancer. *J. Digit. Imag.*, 23(5):611–631, 2010. DOI: 10.1007/s10278-009-9257-x. 39

[126] R. Riffenburgh. *Statistics in Medicine.* Academic Press, 1999. 117

[127] A. Rosenfeld and A. Kak. *Digital Picture Processing*, vol. 2, 2nd ed. Academic Press, 1982. 113, 114

[128] B. Sahiner, H. Chan, N. Petrick, M. Helvie, and M. Goodsitt. Computerized characterization of masses on mammograms: The rubber band straightening transform and texture analysis. *Med. Phys.*, 25:516–526, 1998. DOI: 10.1118/1.598228. 111

[129] B. Sahiner, H. Chan, N. Petrick, M. Helvie, and L. Hadjiiski. Improvement of mammographic mass characterization using spiculation measures and morphological features. *Med. Phys.*, 28:1455–1465, 2001. DOI: 10.1118/1.1381548. 105, 111, 117

[130] M. Sampat, M. Markey, and A. Bovik. Computer-aided detection and diagnosis in mammography. In *Handbook of Image and Video Processing*, pages 1195–1217. Academic Press, New York, 2005. DOI: 10.1016/b978-012119792-6/50130-3. 120

[131] M. Sampat, Z. Wang, S. Gupta, A. Bovik, and M. Markey. Complex wavelet structural similarity: A new image similarity index. *IEEE Trans. Image Process.*, 18(11):2385–2401, 2009. DOI: 10.1109/tip.2009.2025923. 92, 94, 97

[132] M. Sato, M. Kawai, Y. Nishino, D. Shibuya, N. Ohuchi, and T. Ishibashi. Cost-effectiveness analysis for breast cancer screening: Double reading vs. single + CAD reading. *Breast Cancer*, 21(5):532–541, 2014. DOI: 10.1007/s12282-012-0423-5. 13

[133] D. Scutt, G. Lancaster, and J. Manning. Breast asymmetry and predisposition to breast cancer. *Breast Cancer Res.*, 8:R14, 2006. DOI: 10.1186/bcr1388. 7, 76

[134] E. Sickles. Mammography: Asymmetries, masses, and architectural distortion. In J. Hodler, G. von Schulthess, and C. Zollikofer, Eds., *Diseases of the Heart and Chest, Including Breast 2011–2014*, pages 255–258. Springer, 2011. DOI: 10.1007/978-88-470-1938-6. 7, 76

[135] C. Silva, C. Lima, and J. Correia. Breast skin-line detection using dynamic programming. In *Proc. of the Annual International Conference of the IEEE Engineering in Medicine and Biology Society*, pages 7775–7778, 2011. DOI: 10.1109/iembs.2011.6091916. 32, 56, 63, 64, 65

[136] E. Simoncelli, W. Freeman, E. Adelson, and D. Heeger. Shiftable multi-scale transforms. *IEEE Trans. Inf. Theory*, 38(2):587–607, 1992. DOI: 10.1109/18.119725. 92, 94

[137] S. Singletary, C. Allred, P. Ashley, L. Bassett, D. Berry, K. Bland, P. Borgen, G. Clark, S. Edge, D. Hayes, L. Hughes, R. Hutter, M. Morrow, D. Page, A. Recht, R. Theriault, A. Thor, D. Weaver, H. Wieand, and F. Greene. Revision of the american joint committee on cancer staging system for breast cancer. *J. Clin. Oncol.*, 20(17):3628–3636, 2002. DOI: 10.1200/JCO.2002.02.026. 1

[138] S. Singletary, G. Robb, and G. Hortobagyi. *Advanced Therapy of Breast Disease.* BC Decker Inc., 2004. 129

[139] R. Smith, D. Manassaram-Baptiste, D. Brooks, V. Cokkinides, M. Doroshenk, D. Saslow, R. Wender, and O. Brawley. Cancer screening in the United States, 2014: A review of current American Cancer Society guidelines and issues in cancer screening. *CA-Cancer J. Clin.*, 64(1):30–51, 2014. DOI: 10.3322/caac.21174. 76

[140] R. software. http://www.radiology.uchicago.edu/ 15, 86, 97, 101, 110, 116

[141] J. Suckling, J. Parker, D. Dance, S. Astley, I. Hutt, C. Boggis, I. Ricketts, E. Stamakis, N. Cerneaz, S. Kok, P. Taylor, D. Betal, and J. Savage. The mammographic image analysis society digital mammogram database. *Excerpta Medica, International Congress Series 1069*, pages 242–248, 1994. 20, 21, 31, 32, 33, 56, 59, 60, 64, 71, 77, 81, 91, 93, 96, 102, 121, 124, 125

[142] Y. Sun, J. Suri, J. Desautels, and R. Rangayyan. A new approach for breast skin-line estimation in mammograms. *Pattern Anal. Applic.*, 9(1):34–47, 2006. DOI: 10.1007/s10044-006-0023-0. 31, 32, 56, 63, 64, 65

[143] J. Suri, R. Haralick, and F. Sheehan. Greedy algorithm for error correction in automatically produced boundaries from low contrast ventriculograms. *Pattern Anal. Applic.*, 3(1): 39–60, 2000. DOI: 10.1007/s100440050005. 64

[144] L. Tabár, M. Yen, B. Vitak, H. Chen, R. Smith, and S. Duffy. Mammography service screening and mortality in breast cancer patients: 20-year follow-up before and after introduction of screening. *Lancet*, 361:1405–1410, 2003. DOI: 10.1016/s0140-6736(03)13143-1. 1

[145] L. Tabár, T. Tot, and P. Dean. *Breast Cancer, the Art and Science of Early Detection with Mammography: Perception, Interpretation, Histopathologic Correlation*. George Thieme Verlag, 2005. DOI: 10.1055/b-002-59230. 8, 39, 42, 80

[146] M. Tan, B. Zheng, P. Ramalingam, and D. Gur. Prediction of near-term breast cancer risk based on bilateral mammographic feature asymmetry. *Acad. Radiol.*, 20(12):1542–1550, 2013. DOI: 10.1016/j.acra.2013.08.020.

[147] J. Tang, R. Rangayyan, J. Xu, I. E. Naqa, and Y. Yang. Computer-aided detection and diagnosis of breast cancer with mammography: Recent advances. *IEEE Trans. Inf. Technol. Biomed.*, 13(2):236–251, 2009. DOI: 10.1109/titb.2008.2009441. 77, 120

[148] S. Thiruvenkadam, M. Acharyya, N. Neeba, P. Jhunjhunwala, and S. Ranjan. A region-based active contour method for extraction of breast skin-line in mammograms. In *Proc. of the IEEE International Symposium on Biomedical Imaging*, pages 189–192, 2010. DOI: 10.1109/isbi.2010.5490383. 32

[149] J. Thomson, A. Evans, S. Pinder, H. Burrel, A. Wilson, and I. Ellis. Growth pattern of ductal carcinoma in situ (DCIS): A retrospective analysis based on mammographic findings. *Br. J. Cancer*, 85(2):225–227, 2001. DOI: 10.1054/bjoc.2001.1877. 39

[150] S. Tzikopoulos, M. Mavroforakis, H. Georgiou, N. Dimitropoulos, and S. Theodoridis. A fully automated scheme for mammographic segmentation and classification based on breast density and asymmetry. *Comput. Meth. Prog. Bio.*, 102(1):47–63, 2011. DOI: 10.1016/j.cmpb.2010.11.016. 39, 77, 90, 102

[151] L. V. Valen. A study of fluctuating asymmetry. *Evolution*, 16:125–142, 1962. DOI: 10.2307/2406192. 75

[152] S. van Engeland, S. Timp, and N. Karssemeijer. Finding corresponding regions of interest in mediolateral oblique and craniocaudal mammographic views. *Med. Phys.*, 33(9):3203–3212, 2006. DOI: 10.1118/1.2230359. 30, 39, 50, 51, 52

[153] C. Varela, P. Tahoces, A. Méndez, M. Souto, and J. Vidal. Computerized detection of breast masses in digitized mammograms. *Comput. Biol. Med.*, 37(2):214–226, 2007. DOI: 10.1016/j.compbiomed.2005.12.006. 124, 140

[154] A. Venkatesan, P. Chu, K. Kerlikowske, E. Sickles, and R. Smith-Bindman. Positive predictive value of specific mammographic findings according to reader and patient variables. *Radiology*, 250(3):648–657, 2009. DOI: 10.1148/radiol.2503080541. 7, 76

[155] C. H. Waddington. *The Strategy of the Genes*, vol. 20. Routledge, 2014. DOI: 10.4324/9781315765471. 75

[156] X. Wang, D. Lederman, J. Tan, X. Wang, and B. Zheng. Computerized detection of breast tissue asymmetry depicted on bilateral mammograms: A preliminary study of breast risk stratification. *Acad. Radiol.*, 17(10):1234–1241, 2010. DOI: 10.1016/j.acra.2010.05.016. 77, 102

[157] X. Wang, D. Lederman, J. Tan, X. Wang, and B. Zheng. Computerized prediction of risk for developing breast cancer based on bilateral mammographic breast tissue asymmetry. *Med. Eng. Phys.*, 33(8):934–942, 2011. DOI: 10.1016/j.medengphy.2011.03.001. 77, 90, 102

[158] X. Wang, L. Li, W. Xu, W. Liu, D. Lederman, and B. Zheng. Improving performance of computer-aided detection of masses by incorporating bilateral mammographic density asymmetry: An assessment. *Acad. Radiol.*, 19(3):303–310, 2012. DOI: 10.1016/j.acra.2011.10.026. 77

[159] Z. Wang, A. Bovik, H. Sheikh, and E. Simoncelli. Image quality assessment: From error visibility to structural similarity. *IEEE Trans. Image Process.*, 13(4):1–14, 2004. DOI: 10.1109/tip.2003.819861. 92, 97

[160] J. Wei, H. Chan, B. Sahiner, C. Zhou, L. Hadjiiski, M. Roubidoux, and M. Helvie. Computer-aided detection of breast masses on mammograms: Dual system approach with two-view analysis. *Med. Phys.*, 36(10):4451–4460, 2009. DOI: 10.1118/1.3220669. 46

[161] J. Wolfe. Breast parenchymal patterns and their changes with age. *Radiology*, b(121): 545–552, 1976. DOI: 10.1148/121.3.545. 7

[162] Y. Wu, H. Chan, C. Paramagul, L. Hadjiiski, C. Daly, J. Douglas, Y. Zhang, B. Sahiner, J. Shi, and J. Wei. Dynamic multiple thresholding breast boundary detection algorithm for mammograms. *Med. Phys.*, 37(1):391–401, 2010. DOI: 10.1118/1.3273062. 32

[163] R. D. Yapa and K. Harada. Breast skin-line estimation and breast segmentation in mammograms using fast-marching method. *Int. J. Biol. Life Sci.*, 3(1):54–62, 2007. 32

[164] F. Yin, M. Giger, K. Doi, C. Vyborny, and R. Schmidt. Computerized detection of masses in digital mammograms: Automated alignment of breast images and its effect on bilateral subtraction technique. *Med. Phys.*, 21(3):445–452, 1994. DOI: 10.1118/1.597307. 30, 31, 53, 77

[165] Y. Yuan, M. Giger, H. Li, and C. Sennett. Correlative feature analysis on FFDM. *Med. Phys.*, 35(12):5490–5500, 2008. DOI: 10.1118/1.3005641. 39

[166] Y. Yuan, M. Giger, H. Li, N. Bhooshan, and C. Sennett. Correlative analysis of FFDM and DCE-MRI for improved breast CADx. *J. Med. Biol. Eng.*, 32(1):42–50, 2012. DOI: 10.5405/jmbe.833. 39, 42

[167] H. Zhang, P. Heffernan, and L. Tabár. User interface and viewing workflow for mammography workstation. US Patent 2009/0185732 A1, 2009. 80

[168] Z. Zhang, L. Lu, and Y. Yip. Automatic segmentation for breast skin-line. In *Proc. of the 10th IEEE International Conference on Computer and Information Technology*, pages 1599–1604, 2010. DOI: 10.1109/cit.2010.283. 32

[169] B. Zheng, J. Tan, M. Ganott, D. Chough, and D. Gur. Matching breast masses depicted on different views: A comparison of three methods. *Acad. Radiol.*, 16(11):1338–1347, 2009. DOI: 10.1016/j.acra.2009.05.005. 39

[170] B. Zheng, J. Sumkin, M. Zuley, X. Wang, A. Klym, and D. Gur. Bilateral mammographic density asymmetry and breast cancer risk: A preliminary assessment. *Eur. J. Radiol.*, 81 (11):3222–3228, 2012. DOI: 10.1016/j.ejrad.2012.04.018. 7, 76

[171] C. Zhou, H. Chan, C. Paramagul, M. Roubidoux, B. Sahiner, L. Hadjiiski, and N. Petrick. Computerized nipple identification for multiple image analysis in computer-aided diagnosis. *Med. Phys.*, 31(10):2871–2882, 2004. DOI: 10.1118/1.1800713. 30, 53

Authors' Biographies

PAOLA CASTI

Paola Casti graduated with honors in Medical Engineering in 2011 from the University of Rome Tor Vergata, with a thesis titled "Development and validation of a computer-aided detection system for the identification of bilateral asymmetry in mammographic images." In 2015, she received a Ph.D. in Telecommunications and Microelectronics Engineering with the mention of *Excellent Quality Cum Laude*. The title of her dissertation was "Development of an innovative system for early detection and characterization of breast cancer." During her engineering studies, she was a four-time winner of the annual student award of excellence (given to the top 1% of engineering students) from the University of Rome Tor Vergata. In 2009, she was awarded 1st place in a student competition for the mechanical analysis and computer-aided design of a patented prosthesis, with a project titled "Multi-configurable Wrist Joint Prosthesis." She collaborated with the National Institute of Health (Istituto Superiore di Sanitá, ISS) in 2009 on a project for microtomographic evaluation of bone substitutes and the obtained results have been published in the Rapporti ISTISAN of the ISS. In 2012, she was selected as one of the young researcher participants of the IEEE EMBS International Summer School on Biomedical Imaging, held in Île de Berder, Brittany, France. She has coauthored several papers in international peer-reviewed journals, a number of papers in proceedings of international conferences, and technical reports. Her research interests are in signal and image analysis for medical applications, pattern recognition and classification, and computer-aided diagnosis. At present, she is working with the Department of Electronics Engineering of the University of Rome Tor Vergata with a research contract.

ARIANNA MENCATTINI

Arianna Mencattini received an M.S. degree in Electronic Engineering from the University of Rome Tor Vergata, Rome, Italy in 2000, summa cum laude, and a Ph.D. in 2004 for her research on fuzzy logic systems for modelling from the same University. In 2001, she received the award of a Young Researchers Grant and in 2004–2005 she had a scholarship from the Sixth Framework Programme of the European Community at the University of Rome Tor Vergata. Since 2006 she has been an Assistant Professor at the Department of Electronic Engineering, University of Rome Tor Vergata. She is member of the Italian Electrical and Electronic Measurement Group. At present, she teaches a course on image processing. Her main research interests are related to image processing techniques for the development of computed-assisted diagnosis systems, analysis of speech and facial expressions for automatic emotion recognition, and design of novel cell tracking algorithms for immune-cancer interaction analysis. She is the Principal Investigator of the project "PainTCare, Personal pAIn assessemeNT by an enhanCed multimodAl architecture," funded by the University of Rome Tor Vergata, for the automatic assessment of pain in postsurgical patients, and team member of the Project Horizon 2020 "PhasmaFOOD: Portable photonic miniaturised smart system for on-the-spot food quality sensing." She has coauthored more than 70 papers in international journals and conferences.

MARCELLO SALMERI

Marcello Salmeri received a M.Sc. degree in Electronic Engineering and a Ph.D. in Electronic Engineering from the University of Rome Tor Vergata, in 1989 and 1993, respectively. He is currently an Associate Professor with the Department of Electronic Engineering, University of Rome Tor Vergata, and Coordinator of the Electronic Engineering courses. His research interests include signal and image processing; theory, applications, and implementations of fuzzy systems; and pattern recognition. He is the author of about 100 publications in the fields of electronics, measurement, and data analisys.

RANGARAJ MANDAYAM RANGAYYAN

Rangaraj Mandayam Rangayyan is Professor Emeritus of Electrical and Computer Engineering at the University of Calgary, Calgary, Alberta, Canada. He received a Bachelor of Engineering degree in Electronics and Communication Engineering in 1976 from the University of Mysore at the People's Education Society College of Engineering, Mandya, Karnataka, India, and a Ph.D. in Electrical Engineering from the Indian Institute of Science, Bangalore, Karnataka, India, in 1980. He served the University of Manitoba, Winnipeg, Manitoba, Canada and the University of Calgary in research, academic, and administrative positions from 1981–2016. His research interests are in digital signal and image processing, biomedical signal and image analysis, and computer-aided diagnosis. Dr. Rangayyan has published more than 160 papers in journals and 270 papers in proceedings of conferences. He has supervised or cosupervised 27 Master's theses, 17 Doctoral theses, and more than 50 researchers at various levels. He has been recognized with the 1997 and 2001 Research Excellence Awards of the Department of Electrical and Computer Engineering, the 1997 Research Award of the Faculty of Engineering, by appointment as "University Professor" (2003–2013) at the University of Calgary, and with an Outstanding Teaching Performance Award of the Schulich School of Engineering (2016). He is the author of two textbooks: *Biomedical Signal Analysis* (IEEE/ Wiley, 2002, 2015) and *Biomedical Image Analysis* (CRC, 2005). He has coauthored and coedited several other books, including *Color Image Processing with Biomedical Applications* (SPIE, 2011). He has been recognized with the 2013 IEEE Canada Outstanding Engineer Medal, the IEEE Third Millennium Medal (2000), and elected as Fellow, IEEE (2001); Fellow, Engineering Institute of Canada (2002); Fellow, American Institute for Medical and Biological Engineering (2003); Fellow, SPIE (2003); Fellow, Society for Imaging Informatics in Medicine (2007); Fellow, Canadian Medical and Biological Engineering Society (2007); Fellow, Canadian Academy of Engineering (2009); and Fellow, Royal Society of Canada (2016).